George Washington Holley

The Falls of Niagara

The Other Famous Cataracts of the World

George Washington Holley

The Falls of Niagara
The Other Famous Cataracts of the World

ISBN/EAN: 9783743442726

Manufactured in Europe, USA, Canada, Australia, Japa

Cover: Foto ©Andreas Hilbeck / pixelio.de

Manufactured and distributed by brebook publishing software (www.brebook.com)

George Washington Holley

The Falls of Niagara

NIAGARA.

Frontispiece.

Niagara Falls, from the Canadian side.

THE FALLS

OF

NIAGARA

WITH SUPPLEMENTARY CHAPTERS ON

THE OTHER FAMOUS CATARACTS OF THE WORLD.

BY

GEORGE W. HOLLEY.

With Illustrations.

NEW-YORK:
A. C. ARMSTRONG & SON,
714 BROADWAY.
1883.

To the Memory

OF

MY DEPARTED FRIENDS,

PETER A. PORTER
AND
JOEL R. ROBINSON.

The former,

ANSWERING THE SUMMONS TO THE BATTLE-FIELD,
GAVE HIS LIFE FOR HIS COUNTRY.

The latter

OFTEN RISKED HIS OWN LIFE IN SAVING
THE LIVES OF OTHERS.

GEORGE W. HOLLEY.

CONTENTS.

	PAGE
PREFACE........	xiii

PART I.—HISTORY.

CHAPTER I.

First French expedition—Jacques Cartier—He first hears of the great Cataract—Champlain—Route to China—La Salle—Father Hennepin's first and second visits to the Falls................................ 1

CHAPTER II.

Baron La Hontan's description of the Falls—M. Charlevoix's letter to Madame Maintenon—Number of the Falls—Geological indications—Great projection of the rock in Father Hennepin's time—Cave of the Winds—Rainbows............................ 9

CHAPTER III.

The name Niagara—The musical dialect of the Hurons—Niagara one of the oldest of Indian names—Description of the River, the Falls, and the surrounding country.................................. 15

CHAPTER IV.

Niagara a tribal name—Other names given to the tribe—The Niagaras a superior race—The true pronunciation of Indian words..... 19

CHAPTER V.

The lower Niagara — Fort Niagara — Fort Mississauga — Niagara village — Lewiston — Portage around the Falls — The first railroad in the United States — Fort Schlosser — The ambuscade at Devil's Hole — La Salle's vessel, the *Griffin* — The Niagara frontier 25

PART II.—GEOLOGY.

CHAPTER VI.

America the old world — Geologically recent origin of the Falls — Evidence thereof — Captain Williams's surveys for a ship-canal — Former extent of Lake Michigan — Its outlet into the Illinois River — The Niagara Barrier — How broken through — The birth of Niagara...... 32

CHAPTER VII.

Composition of the terrace cut through — Why retrocession is possible — Three sections from Lewiston to the Falls — Devil's Hole — The Medina group — Recession long checked — The Whirlpool — The narrowest part of the river — The mirror — Depth of the water in the Chasm — Former grand Fall 42

CHAPTER VIII.

Recession above the present position of the Falls — The Falls will be higher as they recede — Reason Why — Professor Tyndall's prediction — Present and former accumulations of rock — Terrific power of the elements — Ice and ice bridges — Remarkable geognosy of the lake region 50

PART III.

LOCAL HISTORY AND INCIDENTS.

CHAPTER IX.

Forty years since — Niagara in winter — Frozen spray — Ice foliage and ice apples — Ice moss — Frozen fog — Ice islands — Ice statues — Sleigh-riding on the American Rapids — Boys coasting on them — Ice gorges..... 62

CONTENTS. ix

CHAPTER X. PAGE

Judge Porter — General Porter — Goat Island — Origin of its name — Early dates found cut in the bark of trees and in the rock — Professor Kalm's wonderful story — Bridges to the Island — Method of construction — Red Jacket — Anecdotes — Grand Island — Major Noah and the New Jerusalem — The Stone Tower — The Biddle stairs — Sam Patch — Depth of water on the Horseshoe — Ships sent over the Falls 71

CHAPTER XI.

Joel R. Robinson, the first and last navigator of the Rapids — Rescue of Chapin — Rescue of Allen — He takes the *Maid of the Mist* through the Whirlpool — His companions — Effect upon Robinson — Biographical notice — His grave unmarked 85

CHAPTER XII.

A fisherman and a bear in a canoe — Frightful experience with floating ice — Early farming on the Niagara — Fruit-growing — The original forest — Testimony of the trees — The first hotel — General Whitney — Cataract House — Distinguished visitors — Carriage road down the Canadian bank — Ontario House — Clifton House — The Museum — Table and Termination Rocks — Burning Spring — Lundy's Lane — Battle Anecdotes 96

CHAPTER XIII.

Incidents — Fall of Table Rock — Remarkable phenomenon in the river — Driving and lumbering on the Rapids — Points of the compass at the Falls — A first view of the Falls commonly disappointing — Lunar bow — Golden spray — Gull Island and the gulls — The highest water ever known at the Falls — The Hermit of the Falls 108

CHAPTER XIV.

Avery's descent of the Falls — The fatal practical joke — Death of Miss Rugg — Swans — Eagles — Crows — Ducks over the Falls — Why dogs have survived the descent........ 118

CHAPTER XV.

Wedding tourists at the Falls — Bridges to the Moss Islands — Railway at the Ferry — List of persons who have been carried over the Falls — Other accidents 125

CHAPTER XVI.

The first Suspension Bridge—The Railway Suspension Bridge—Extraordinary vibration given to the Railway Bridge by the fall of a mass of rock—De Veaux College—The Lewiston Suspension Bridge—The Suspension Bridge at the Falls..................................... 137

CHAPTER XVII.

Blondin and his "ascensions"—Visit of the Prince of Wales—Grand illumination of the Falls—The steamer *Caroline*—The Water-power of Niagara—Lord Dufferin and the plan of an international park.... 144

CHAPTER XVIII.

Poetry in the Table Rock albums—Poems by Colonel Porter, Willis G. Clark, Lord Morpeth, José Maria Heredia, A. S. Ridgely, Mrs. Sigourney, and J. G. C. Brainard.............................. 153

PART IV.

OTHER FAMOUS CATARACTS OF THE WORLD.

CHAPTER XIX.

Yosemite—Vernal—Nevada—Yellowstone—Shoshone—St. Maurice—Montmorency............................. 164

CHAPTER XX.

Tequendama—Kaieteur—Paulo Affonso—Keel-fos—Riunkan-fos—Sarp-fos—Staubbach—Zambesi or Victoria—Murchison—Cavery—Schaffhausen 171

CHAPTER XXI.

Famous rapids and cascades—Niagara—Amazon—Orinoco—Parana—Nile—Livingstone............................. 179

LIST OF ILLUSTRATIONS.

Niagara Falls from the Canadian Side..	....	.Frontispiece.	
The Horseshoe Fall from Goat Island.	..	Opposite page	6
Luna Fall and Island in Winter......		" "	11
The Rapids above the Falls.............	....	" "	17
The Youngest Inhabitant......		" "	22
Mouth of the Chasm and Brock's Monument....		" "	29
Niagara Falls from Below...		" "	54
Great Icicles under the American Fall		" "	60
Winter Foliage		" "	66
Ice Bridge and Frost Freaks		" "	69
Coasting below the American Fall.........		" "	70
Second Moss Island Bridge. ...		" "	76
Joel R. Robinson...................		" "	86
The *Maid of the Mist* in the Whirlpool.		" "	91
Fisher and the Bear............		" "	97
Fall of Table Rock...............		" "	109
Rock of Ages and Whirlwind Bridge		" "	114
The Three Sisters or Moss Islands ...		" "	125
How the Suspension Bridge was Begun..		" "	137

LIST OF ILLUSTRATIONS.

Blondin Crossing the Niagara	Opposite page	145
Indian Women Selling Bead-work	" "	148
Yosemite Falls	" "	164
Bridal Veil Fall	" "	166
Vernal Falls	" "	168
Nevada Falls	" "	171
Lower Falls of the Yellowstone	" "	172
Upper Falls of the Yellowstone	" "	174
The Staubbach, Switzerland	" "	176
Victoria Falls, Zambesi	" "	178
Map of the Niagara Region	" "	1

PREFACE.

THE writer, having resided in the village of Niagara Falls for more than a third of a century, has had opportunity to become thoroughly acquainted with the locality, and to study it with constantly increasing interest and admiration. Long observation enables him to offer some new suggestions in regard to the geological age of the Falls, their retrocession, and the causes which have been potent in producing it; and also to demonstrate the existence of a barrier or dam that was once the shore of an immense fresh-water sea, which reached from Niagara to Lake Michigan, and emptied its waters into the Gulf of Mexico.

Whoever undertakes to write comprehensively on this subject will soon become aware of the weakness of exclamation points and adjectives, and the almost irresistible temptation to indulge in a style of composition which he cannot maintain, and should not if he could. So far as the writer, yielding to the inspiration of his theme, and in opposition to all resolutions to the contrary, may have trespassed in this direction, he bares and bows his head to the severest treatment that the critic may adopt. His labor has been one of love, and in giving its results to the public he regrets that it is not more worthy of the subject.

As it is hoped that the work may be useful to future visitors to the Falls, and also possess some interest for those who have visited them, it seemed desirable to avoid the introduction of notes and the citation of authorities. For this reason several paragraphs are placed in the text which would otherwise have been introduced in notes. This is especially true of the chapters of local history.

The writer is especially indebted to the Hon. Orsamus H. Marshall, of Buffalo, for a copy of his admirable "Historical Sketches," and for access to his library of American history. The Documentary History and Colonial Documents of the State of New York, "The Relations of the Jesuits," the works of other early French missionaries, travelers, and adventurers, made familiar to the public by the indefatigable labors of Shea and Parkman, have all helped to make the writer's task comparatively an easy one.

PREFACE.

Several years ago, the body of this work, which has since been revised and considerably enlarged, was published in a small volume, that has long been out of print. Believing that the interest of the volume would be enhanced for the reader if he were able to contrast Niagara Falls with other famous falls, cataracts, and rapids, the writer has added chapters, describing the most noted of these in all parts of the world.

G. W. H.

NIAGARA FALLS, N. Y.
September, 1882.

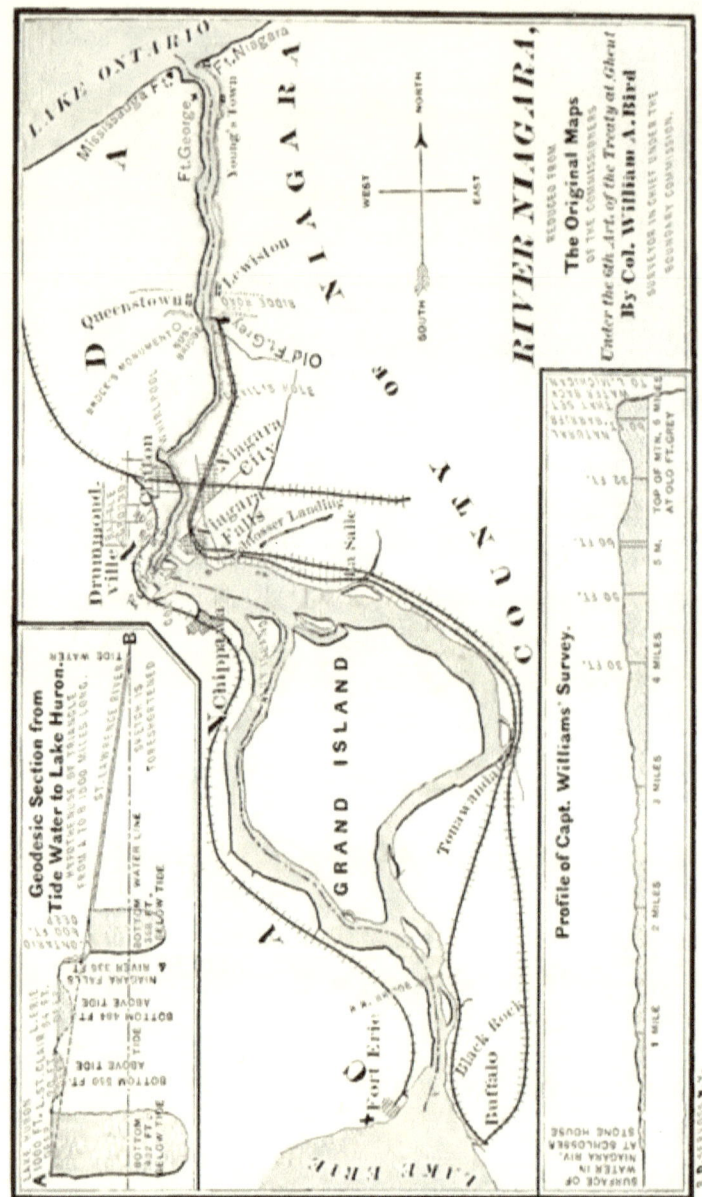

Opposite page 1.

PART I.—HISTORY.

CHAPTER I.

First French expedition — Jacques Cartier — He first hears of the great Cataract — Champlain — Route to China — La Salle — Father Hennepin's first and second visits to the Falls.

IN 1534, Jacques Cartier, a shrewd, enterprising, and adventurous sailor, made his first voyage across the Atlantic, touching at Newfoundland, and exploring the coast to the west and south of it. The two vessels of Cartier, called ships by the historians of the period, were each of only forty tons burden.

On the return of Cartier to France, so favorable was his report of the results of the expedition, that Francis I. commissioned him, the year following, for another voyage, and in May, 1535, after impressive religious ceremonies, he sailed with three vessels thoroughly equipped. The record of this second voyage of Cartier, by Lescarbot, contains the first historical notice of the cataract of Niagara. The navigator, in answer to his inquiries concerning the source of the St. Lawrence, " was told that, after ascending many leagues among rapids and waterfalls, he would reach a lake one hundred and forty or fifty leagues broad, at the western extremity of which the waters were wholesome and the winters mild; that a river emptied into it from the south, which had its source in the country of the Iroquois; that beyond the lake he would find a cataract and portage, then another lake about equal to the former, which they had never explored."

In 1603, a company of merchants in Rouen obtained the necessary authority for a new expedition to the St. Lawrence, which they placed under the direction of Samuel Champlain, an able, discreet, and resolute commander. On a map published in 1613 he indicated the position of the cataract, calling it merely a water-fall *(saut d'eau)*, and describing it as being "so very high that many kinds of fish are stunned in its descent." It does not appear by the record that he ever saw the Falls.

During the sixty years that elapsed between the establishment of the French settlements by Champlain and the expedition of La Salle and Hennepin, there can be little doubt that the great cataract was repeatedly visited by French traders and adventurers. Many of the earlier travelers to the region of the St. Lawrence believed that China could be reached by an overland journey across the northern part of the continent. Father Vimont informs us (" Relations of the Jesuits," 1642-3) that the Jesuit Raymbault "designed to go to China across the American wilderness, but God sent him on the road to heaven." As he died at the Saut Ste.

Marie in 1641, he must have passed to the north of the Falls without seeing them. In 1648, the Jesuit father Ragueneau, in a letter to the Superior of the Mission, at Paris, says: "North of the Eries is a great lake, about two hundred leagues in circumference, called Erie, formed by the discharge of the *mer-douce* or Lake Huron, and which falls into a third lake, called Ontario, over a cataract of frightful height."

In some important manuscripts relating to the earliest expeditions of the French into Canada,—discovered a few years ago, and now in the possession of M. Pierre Margry, of Paris,—occurs a description of the Falls communicated by the Indians to Father Gallinée, one of the two Sulpician priests who accompanied La Salle in his first visit to the Senecas, in 1669. He seems to have been more indifferent to the charms of Nature than Father Raymbault, since he crossed the Niagara River near its mouth, and within hearing of its falling waters, yet did not turn aside to see the cataract. In his journal he says: "We found a river one-eighth of a league broad and extremely rapid, forming the outlet of Lake Erie and emptying into Lake Ontario. The depth of the river is, at this place, extraordinary, for, on sounding close by the shore, we found fifteen or sixteen fathoms of water. This outlet (the Niagara River) is forty leagues long, and has, from ten to twelve leagues above Lake Ontario, one of the finest cataracts in the world; for all the Indians of whom I have inquired about it say that the river falls at that place from a rock higher than the tallest pines—that is, about two hundred feet. In fact, we heard it from the place where we were, although from ten to twelve leagues distant, but the fall gives such a momentum to the water that its velocity prevented our ascending the current by rowing, except with great difficulty. At a quarter of a league from the outlet, where we were, it grows narrower, and its channel is confined between two very high, steep, rocky banks, inducing the belief that the navigation would be very difficult quite up to the cataract. As to the river above the Falls, the current very often sucks into this gulf, from a great distance above, deer and stags, elk and roebucks, which, in attempting to swim the river, suffer themselves to be drawn so far down-stream that they are compelled to descend the Falls, and are overwhelmed in its frightful abyss.

"Our desire to reach the little village called Ganastoque Sonontona (between the west end of Lake Ontario and Grand River) prevented our going to view that wonder. * * * I will leave you to judge if that must not be a fine cataract, in which all the water of the large river (St. Lawrence) * * * falls from a height of two hundred feet, with a noise that is heard not only at the place where we were,—ten or twelve leagues distant,— but also from the other side of Lake Ontario, opposite its mouth" (Toronto, forty miles distant).

Of the rattlesnakes on the mountain ridges he says: "There are many in this place as large as your arm, and six or seven feet long, and entirely black."

From Ganastoque Sonontona the party separated, the two priests, with their guides and attendants, designing to move to the west, along the north shore of Lake Erie, and La Salle apparently to return to Montreal, but in reality, as is supposed, to prosecute by a more southerly route the grand ambition of his life—the discovery of the Mississippi River—a purpose which he executed with even more than the "bigot's zeal," and literally, as it proved in the end, with the "martyr's constancy," for he was assassinated on the plains of Texas, some few years after, while endeavoring to secure to France the benefits of his great discovery.

After separating from his companions at the Indian village, he probably returned to Lake Ontario and the Niagara River, which he crossed, no doubt, on his way to some of the Iroquois villages, in search of a guide and attendants to assist him in his explorations. It may be assumed that he visited the Falls at this time, but his journal of this expedition has never been found.

The first description of the Falls by an eye-witness is that of Father Hennepin, so well known to those conversant with our early history. He saw it for the first time in the winter of 1678-9, and his exaggerated account of it is accompanied by a sketch which in its principal features is undoubtedly correct, though its perspective and proportions are quite otherwise. He says: "Betwixt the lakes Ontario and Erie there is a vast and prodigious cadence of water, which falls down in a surprising and astonishing manner, insomuch that the universe does not afford its parallel. 'Tis true that Italy and Switzerland boast of some such things, but we may well say they are sorry patterns when compared with this of which we now speak. * * * It [the river] is so rapid above the descent, that it violently hurries down the wild beasts while endeavoring to pass it, * * * they not being able to withstand the force of its current, which inevitably casts them headlong above six hundred feet high. This wonderful downfall is composed of two great streams of water and two falls, with an isle sloping along the middle of it. The waters which fall from this horrible precipice do foam and boil after the most hideous manner imaginable, making an outrageous noise, more terrible than that of thunder; for, when the wind blows out of the south, their dismal roaring may be heard more than fifteen leagues off.

"The river Niagara having thrown itself down this incredible precipice, continues its impetuous course for two leagues together to the great rock, above mentioned [in another chapter as lying at the foot of the mountain at Lewiston], with inexpressible rapidity. * * * From the great Fall unto this rock, which is to the west of the river, the two brinks of it are so prodigiously high, that it would make one tremble to look steadily upon the water rolling along with a rapidity not to be imagined."

On his return from the West, in the summer of 1681, the Father informs us that he "spent half a day in considering the wonders of that prodigious cascade." Referring to the spray, he says: "The rebounding of these waters is so great that a sort of cloud arises from the foam of it, which is seen hanging over this abyss even at noon-day." Of the river, he says: "From the mouth of Lake Erie to the Falls are reckoned six leagues. * * *

The Horseshoe Fall, from Goat Island.
Opposite page 6.

The lands which lie on both sides of it to the east and west are all level from the Lake Erie to the great Fall." At the end of the six leagues "it meets with a small sloping island, about half a quarter of a league long and near three hundred feet broad, as well as one can guess by the eye. From the end, then, of this island it is that these two great falls of water, as also the third, throw themselves, after a most surprising manner, down into the dreadful gulph, six hundred feet and more in depth." On the Canadian side, he says: "One may go down as far as the bottom of this terrible gulph. The author of this discovery was down there, the more narrowly to observe the fall of these prodigious cascades. From there we could discover a spot of ground which lay under the fall of water which is to the east [American Fall] big enough for four coaches to drive abreast without being wet; but because the ground * * * where the first fall empties itself into the gulph is very steep and almost perpendicular, it is impossible for a man to get down on that side, into the place where the four coaches may go abreast, or to make his way through such a quantity of water as falls toward the gulph, so that it is very probable that to this dry place it is that the rattlesnakes retire, by certain passages which they find underground."

Finding no Indians living at the Falls, he suggests a probable reason therefor: "I have often heard talk of the Cataracts of the Nile, which make people deaf that live near them. I know not if the Iroquois who formerly lived near this fall * * * withdrew themselves from its neighborhood lest they should likewise become deaf, or out of the continual fear they were in of the rattlesnakes, which are very common in this place. * * * Be it as it will, these dangerous creatures are to be met with as far as the Lake Frontenac [Ontario], on the south side; and it is reasonable to presume that the horrid noise of the Fall and the fear of these poisonous serpents might oblige the savages to seek out a more commodious habitation." In the view of the Falls accompanying his description, a large rock is represented as standing on the edge of the Table Rock. This rock is mentioned by Kalm, a Swedish naturalist, who visited the Falls in 1750, as having disappeared a few years before that date. Father Hennepin's reference to the animals drawn into the current and going over the Falls, and to the rattlesnakes, indicates unmistakably his previous acquaintance with Father Gallinées's narrative.

CHAPTER II.

Baron La Hontan's description of the Falls — M. Charlevoix's letter to Madame Maintenon — Number of the Falls — Geological indications — Great projection of the rock in Father Hennepin's time — Cave of the Winds — Rainbows.

EVEN more exaggerated than Father Hennepin's is the next account of the Falls which has come down to us, and which was written by Baron La Hontan, in the autumn of 1687. Fear of an attack from the Iroquois, the relentless enemies of the French, made his visit short and unsatisfactory. He says: "As for the water-fall of Niagara, 'tis seven or eight hundred feet high, and half a league wide. Toward the middle of it we descry an island, that leans toward the precipice, as if it were ready to fall." Concerning the beasts and fish drawn over the precipice, he says they "serve for food" for the Iroquois, who "take 'em out of the water with their canoes"; and also that "between the surface of the water, that shelves off prodigiously, and the foot of the precipice, three men may cross in abreast, without further damage than a sprinkling of some few drops of water." Father Hennepin, it will be remembered, makes this space broad enough for four coaches, instead of three men.

From the Baron's declaration as to the manner in which the Indians captured the game which went over the Falls, it would seem that the bark canoe of the Indian was the precursor of the white man's skiff and yawl, that serve as a ferry below the Falls. And the timid traveler of the present day, who hesitates about crossing in this latter craft, will probably pronounce the Indian foolhardy for venturing on those turbulent waters in his light canoe, whereas, in skillful hands, it is peculiarly fitted for such navigation.

A more correct estimate of the cataract than either of the preceding is that of M. Charlevoix, sent to Madame Maintenon, in 1721. After referring to the inaccurate accounts of Hennepin and La Hontan, he says: "For my own part, after having examined it on all sides, where it could be viewed to the greatest advantage, I am inclined to think we cannot allow it [the height] less than one hundred and forty or fifty feet." As to its figure, "it is in the shape of a horseshoe, and it is about four hundred paces in circumference. It is divided in two exactly in the center by a very narrow island, half a quarter of a league long." In relation to the noise of the falling water, he says: "You can scarce hear it at M. de Joncaire's [Fort Schlosser], and what you hear in this place [Lewiston] may possibly be the whirlpools, caused by the rocks which fill up the bed of the river as far as this."

Neither Baron La Hontan nor M. Charlevoix speaks of the number of water-falls. But Father Hennepin, it will be remembered, mentions three; two of which were to the south and west of Goat Island. And the Rev. Abbé Picquet, who visited the place in 1751,

Luna Fall and Island in Winter.
Opposite page 11.

seventy years after Father Hennepin, says (Documentary History, I., p. 283): "This cascade is as prodigious by reason of its height and the quantity of water which falls there, as on account of the variety of its falls, which are to the number of six principal ones divided by a small island, leaving three to the north and three to the south. They produce of themselves a singular symmetry and wonderful effect."

The geological indications are that Goat Island once embraced all the small islands lying near it, and also that it covered the whole of the rocky bar which stretches up stream some hundred and fifty rods above the head of the present island. At that period, from the depressions now visible in the rocky bed of the river, it would seem probable that the water cut channels through the modern drift corresponding with these depressions. In that case there would then have been a third fall in the American channel, north of Goat Island, lying between Luna Island and a small island then lying just north of the Little Horseshoe, and stretching up toward Chapin's Island. On the south side of Goat Island, there would have been a fall between its southern shore and an island then situated about two hundred feet farther south.

The highest point in the American Fall, the salient and beautiful projection near the shore at Prospect Park, is upheld by a more substantial foundation than is revealed at any other accessible portion of the face of the precipice. This is made manifest on entering the "Shadow-of-the-Rock," where the spectator sees a massive wall of thoroughly indurated limestone, disposed in regular layers more than two feet in thickness, with faces as smooth as if dressed with the chisel. Passing in front of this, across the American Fall, under the Horseshoe and Table Rock, there must have been formerly a broad cleft of soft, friable limestone, to the disintegration and removal of which was due the great overhanging of the upper strata noticed by Father Hennepin and Baron La Hontan.

For three miles above the Falls, the course of the river is almost due west. But after leaving the precipice it makes an acute angle with its former direction, and thence runs north-east to the railway suspension bridge. The formation of the rapids — one of the most beautiful features of the scene — is due to this change of direction. At no point below its present position could there have been such a prelude — musical as well as motional — to the great cataract. And when these rapids shall have disappeared in the receding flood it is not probable that there will be other rapids that can equal them in length, breadth, beauty, and power.

The declivity in the lower channel through the gorge is ninety feet; but on the surface of the upper banks there is a rise of more than one hundred feet in the same direction — that is, down the river. Hence, when the Falls were at Lewiston they were more than two hundred and fifty feet high. Now the greatest descent is one hundred and sixty-eight feet, the diminution being the result of retrocession in the line of the dip — from north-east to south-west — in the bed-rock. It is owing to this dip that the surface of the water on the American side is ten feet higher than it is on the Canadian. The continuous column of water, however, is longest in the center of the Horseshoe, because of the fallen rock and *débris* lying at the foot of the other portions of the Fall. At this time the upward slope of the bed-rock is such that— if it shall prove to be sufficiently hard—the Falls, after receding four miles farther, will be two hundred and twenty feet high.

It is evident from the descriptions of Father Hennepin and of Baron La Hontan, that the upper stratum of rock over which the water falls must have projected beyond the face of the rock below much farther than it now does. The large masses of fallen rock lying at the foot of the American and Horse-shoe Falls are evidence of this fact. Travelers still go behind the sheet on the Canadian side, and into and through the Cave of the Winds, on the American side. But they do not expect to keep dry in so doing, nor to sun themselves on the rocks below, like the "rattlesnakes" of former days. Nevertheless, there is no more exciting nor exhilarating excursion to be made at the Falls than that through the Cave of the Winds.

Nowhere else are the prismatic hues exhibited in such wonderful variety, nor in such surpassing brilliancy and beauty. And although a rainbow is not a spraybow, it may be admitted that a spraybow is a rainbow, formed of drops of water, large or small. So here rainbow dust and shattered rainbows are scattered around; rainbow bars and arches, horizontal and perpendicular, are flashing and forming, breaking and reforming, around and above the visitor in the most fantastic and delightful confusion of form and effect. And if his fancy prompts him, he may arrange himself as a portrait, at half or full length, in an annular bow. The enamored Strephon may literally place his charming Delia in a living, sparkling rainbow-frame, flecked all over with diamonds and pearls.

CHAPTER III.

The name Niagara—The musical dialect of the Hurons—Niagara one of the oldest of Indian names—Description of the river, the Falls, and the surrounding country.

THERE is in some words a mystic power which it is not easy to analyze or define; they fascinate the ear even of those who do not understand their meaning. The very sound of them as they are enunciated by the human voice touches a chord to which the heart instinctively responds. So it is with the name of the great cataract. No one can hear it correctly pronounced without being charmed with its rhythmical beauty, or without feeling confident of its poetical aptness and significance in the dialect from which it was derived.

And although we have no means of determining the correctness of any of the fanciful or poetical interpretations which have been given of the word, still we cannot doubt that it must have had a peculiar force and justness with those who first applied it. Baron La Hontan, who spent several years among the Indians, noticed the remarkable fact concerning their language that it had no labials. "Nevertheless," he says, "the language of the Hurons appears very beautiful, and the sound of it perfectly charming, although, in speaking it, they never close their lips."

The most voluminous and among the earliest existing records connected with the River St. Lawrence, and the great lakes which it drains, are the well-known "Relations of the Jesuits," so called, comprising a yearly account of the labors of the Missionary Fathers sent out by the College at Paris to Christianize the Indians. In 1615, they established their mission at Quebec, and from thence extended their operations westward. In 1626, they reached the large and powerful tribe of Indians which occupied the splendid domain which may be described with proximate accuracy as bounded by a line commencing at a point on the southerly shore of Lake Ontario, about thirty miles west of the mouth of the Genesee River, and running thence parallel to that river to a point due west from Avon; thence nearly due west to Buffalo; thence along the north shore of Lake Erie to the Detroit River; thence up that river to a point directly west from the west end of Lake Ontario; thence east to that lake, and finally along the southern shore of it to the place of beginning.

The oldest and most notable name in all this territory is NIAGARA, as would naturally be inferred, when we consider the varied and wonderful features of the mighty river which flows across this country. Taking leave of Lake Erie, its clear waters gradually spread themselves out in a broad, bright channel, over a plain, open country, having a slight declivity, just sufficient to make a gentle current, thereby adding the living beauty and force of motion to the broad expanse of a lake-like surface, that surface itself diversified and relieved by the pleasant islands, large

The Rapids above the Falls.
Opposite page 17.

and small, which are scattered over it. Eddying into every quiet bay, coquetting with every salient angle, moving to the melody of its own murmurs, it flows on serenely and musically.

But after a time this holiday journey is interrupted. A fearful change takes place. The careless waters are hurried down a long and sharp descent, over the rough, denuded, bowlder-studded bed-rock of the stream. Breaking and bounding, surging and resurging, flashing and foaming, rushing fiercely upon some huge bowlder, recoiling an instant, then madly leaping entirely over it, rushing on to others huger still, then breaking wildly around them, the troubled waters hurry on until, culminating in their sublimest aspect, they plunge sheer downward in the grandest of cataracts.

And now the scene and the effect it produces on the beholder both change. The rapids are beautiful; the falls are grand; those are exhilarating, these are inspiring; those are noisy, turbulent, fickle; these are calm, resistless, inexorable.

After the water has made the final plunge over the precipice the cataract acquires its most impressive characteristics; the majestic monotone, the bow, the cloud, which is its veil by night, its crowning glory and beauty by day. The combinations of grandeur and beauty have reached their climax in the fall, the foam, the voice, the spray, the bow.

The chasm of the river from the Falls to Lewiston will be sufficiently described in treating of the geology of the district. From Lewiston to Lake Ontario, seven miles, the waters of the river flow on through an elevated and fertile plain, in a strong, calm, majestic current, smiling with dimples and reversed in occasional eddies, but neither broken by rapids nor impeded by islands. Finally it is lost in the lake, after passing an immense bar formed by the enormous mass of sedimentary matter carried down by its own current. The landscape, as seen from the top of the terrace above Lewiston, is one of the finest and most extensive of its peculiar character which can be found on the continent, all its features being such as appertain to a broad, open country.

The visitor at Niagara, as he looks at the Falls, will have a profounder appreciation of their magnitude by considering that it requires the water drainage of a quarter of a continent to sustain them, and that the remoter springs, which send to them their constant tribute, are more than twelve hundred miles distant.

CHAPTER IV.

Niagara a tribal name—Other names given to the tribe—The Niagaras a superior race—The true pronunciation of Indian words.

THE name Niagara has been so thoroughly identified with the river and the Falls that the question whether it was also the name of an Indian nation or tribe has been quite neglected. It is proposed now to give the question some consideration, assuming, at once, its affirmative to be true. This, it is believed, we shall be justified in doing by every principle of analogy. We know that it was a general practice of the Indians who occupied this region of country, so abounding in lakes and rivers, to give the name of the nation or tribe to, or to name them after, the most prominent bodies and courses of water found in their territory. Such was the fact with the Senecas, Cayugas, Oneidas, Onondagas, and Hurons, the tribal name of each being perpetuated both in a lake and a river. The Mohawks, the warrior tribe of the Six Nations, having no noted lake within their boundaries, left a perpetual memorial of themselves in the name of a beautiful river. The unwarlike Eries, too, though finally exterminated by their more powerful and aggressive neighbors, the Iroquois, are still remembered in the lake which bears their name.

With the Niagaras the river and the cataract were the most notable and impressive features of their territory. Their principal village bore the same name; and when we recall the proverbial vanity of the race, we can hardly doubt that this must also have been their tribal name. That it should have been perpetuated in reference to the village, the river, and the falls, and that the use of it, in reference to the tribe, should have lapsed, can be readily understood when we recollect that they had two substitutes for the tribal name. One of these substitutes is explained at page 70 of the "Relations" of 1641, in a passage which we translate as follows: "Our Hurons call the Neuter Nation *Attouanderonks*, as though they would say a people of a little different language: for as to those nations that speak a language of which they understand nothing, they call them *Attouankes*, whatever nation they may be, or as though they spoke of strangers. They of the Neuter Nation in turn, and for the same reason, call our Hurons *Attouanderonks*."

Thus it would seem that this was a mere title of convenience used to indicate a certain fact, namely, a difference of language. The other substitute by which the nation was best known among their white brethren will be understood by an extract from a letter contained in the same "Relations," and written from St. Mary's Mission on the river Severn, by Father Lalement. In it he gives an account of a journey made by the Fathers Jean de Brebeuf and Joseph Marie Chaumont to the country of the *Neuter Nation*, as the Niagaras were called by the

Hurons on the north and the Iroquois on the south of them, learning it, as they did, from the French. The letter says: "Our French, who first discovered this people, named them the *Neuter Nation*, and not without reason, for their country being the ordinary passage by land, between some of the Iroquois nations and the Hurons, who are sworn enemies, they remained at peace with both; so that in times past the Hurons and the Iroquois, meeting in the same wigwam or village of that nation, were both in safety while they remained. There are some things in which they differ from our Hurons. They are larger, stronger, and better formed. They also entertain a great affection for the dead. * * * The Sonontonheronons [Senecas], one of the Iroquois nations the nearest to and most dreaded by the Hurons, are not more than a day's journey distant from the easternmost village of the Neuter Nation, named Onguiaahra [Niagara], of the same name as the river."

It would seem, then, that this name, Neuter Nation, as applied to this tribe, was an appellation used merely to indicate a peculiarity of its location, or of the relation in which it stood to the hostile tribes living to the north and south of it. The Indians, it is needless to say, were not philologists, and seem not to have objected to the names applied to them, nor to have criticised the erroneous pronunciation of words of their own dialects.

In the extract given above, the name of our river first appears in type. Its orthography will be noted as peculiar. It is one of forty different ways of spelling the name, thirty-nine of which are given in the index volume of the Colonial History of New York, and the fortieth, the most pertinent to our present purpose, in Drake's "Book of the Indians," seventh edition. Prefixed to "Book First" is a "Table of the Principal Tribes," in which we find the following:

"Nicariagas, once about Michilimakinak; joined the Iroquois in 1723."

M. Charlevoix, apparently using the facts stated in one of Lalement's letters and quoting also a portion of its language, says: "A people larger, stronger, and better formed than any other savages, and who lived south of the Huron country, were visited by the Jesuits, who preached to them the Kingdom of God. They were called the Neuter Nation, because they took no part in the wars which desolated the country. But in the end they could not themselves escape entire destruction. To avoid the fury of the Iroquois, they finally joined them against the Hurons, but gained nothing by the union." Later, he says they were destroyed about the year 1643. But we have before observed that Father Raugeneau states that their destruction occurred in 1651. The tribe mentioned by Drake was probably a remnant that escaped in the final overthrow of their nation in this last-named year, and sought refuge at Mackinaw, among the Hurons, who had previously retreated to this almost inaccessible locality, in order, also, to escape from the all-conquering Iroquois. After the lapse of nearly three-quarters of a century, when the hostility of the latter had subsided, and they had themselves been weakened and

Opposite page 22.   The Youngest Inhabitant.

subdued by the whites, the wretched remnant of the Niagaras, with that strong love of home so characteristic of the Indian, returned to their native hunting-grounds, where they remained for a few years, and then joined their conquerors in that mournful procession of their race toward the setting sun. If there were a Nemesis for nations as well as for individuals, it would be fearful to contemplate the time when the Anglo-Saxon should be called on to pay the "long arrears" of the Indians' "bloody debt."

Returning to the orthography of our name, we find on Sanson's map of Canada, published in Paris in 1657, that it is shortened into "Oniagra," and on Coronelli's map of the same region, published in Paris in 1688, it crystallizes into *Niagara*. There is also on this map a village located on or near the site of Buffalo, designated as follows: "*Kah-kou-a-go-gah, a destroyed nation.*" This name bears a closer resemblance to the true one than several of the forty to which we have just referred, and if it be reduced to Kahkwa it would still be only a corrupt abbreviation of Niagara.

More than fifty years ago, while leisurely traveling through western New York, the writer well remembers how his youthful ears were charmed with the flowing cadences of the better class of Indians, as they intoned rather than spoke the beautiful names which their ancestors had given to different localities. Every vowel was fully sounded.

O-N-E-I-D-A was then Oh-ne-i-dah; C-A-Y-U-G-A was Kah-yu-gah; G-E-N-E-S-E-E was Gen-e-se-e;

C-A-N-A-N-D-A-I-G-U-A was Kan-nan-dar-quah, and N-I-A-G-A-R-A was Ni-ah-gah-rah.

In regard to the name, the pronunciation nearest to the original which it may be possible to perpetuate is Ni-ag-a-rah; the accent on the second syllable, the vowel in the first pronounced as in the word *nigh;* the *a* in the third and fourth syllables but slightly abbreviated from the long *a* in *far*, and that in the second syllable but slightly aspirated.

CHAPTER V.

The lower Niagara—Fort Niagara—Fort Mississauga—Niagara Village—Lewiston—Portage around the Falls—The first railroad in the United States—Fort Schlosser—The ambuscade at Devil's Hole—La Salle's vessel, the *Griffin*—The Niagara frontier.

FROM the earliest visit of the French missionaries and *voyageurs* to the lake region, the banks of the lower Niagara were to them a favorite locality. Very early they were cleared of the grand forest which covered them, and the genial, fertile, and easily worked soil, enriched by the deep vegetable mold that had been accumulating upon it for centuries, produced in lavish abundance wheat, maize, garden vegetables, and fruits, large and small. "On the 6th day of December, 1678," says Marshall, " La Salle, in his brigantine of ten tons, doubled the point where Fort Niagara now stands, and anchored in the sheltered waters of the river. The prosecution of his bold enterprise at that inclement season, involving the exploration of a vast and unknown country, in vessels built on the way, indicates the indomitable energy and self-reliance of the intrepid discoverer. His crew consisted of sixteen persons, under the immediate command of the Sieur de la Motte. The grateful Franciscans chanted '*Te Deum laudamus*' as they entered the noble river. The strains of that ancient hymn of the Church, as they rose from the deck of the adventurous bark, and echoed from shore and forest, must have startled the watchful Senecas with the unusual sound, as they gazed upon their strange visitors. Never before had white men, so far as history tells us, ascended the river."

La Salle rested here for a time, but no defensive work was constructed until 1687, when the Marquis De Nonville, returning from his famous expedition against the Senecas, fortified it, after the fashion of the time, with palisades and ditches. The small garrison of one hundred men which he left were obliged to abandon it the following season, after partially destroying it. By consent of the Iroquois it was reconstructed in stone in 1725-6.

Opposite to Fort Niagara, which is on the American side at the mouth of the river, are Fort Mississauga and the village of Niagara, formerly Newark, on the Canadian side. The village was captured by the English in 1759, and occupied for a time by Sir William Johnson, who completed here his treaty with the Indians by which they released to him the land on both sides of the river. The first Provincial Parliament was held here in 1792, under the authority of Lieutenant-Governor Simcoe. In the same year the place was visited by the father of Queen Victoria. The pioneer newspaper of the Province was published here in 1795, and although it ceased soon after to be the seat of government, which was removed to York (now Toronto), still it was a thriving village of about five thousand inhabitants until the completion of the Welland canal, which entirely diverted its trade and commerce, and left it to the uninterrupted quiet of a rural town. Several Americans have purchased dwellings in the place for summer occupation. A mile above was Fort George, now a ruin.

Seven miles above the mouth of the river, at the head of navigation, nestling at the foot of the so-called mountain, is Lewiston, named in 1805 in honor of Governor Lewis, of New York. Here, in 1678, La Salle "constructed a cabin of palisades to serve as a magazine or storehouse." And this was the commencement of the portage to the river above the Falls, which passed over nearly the same route as the present road from Lewiston, which is still called the Portage Road. Here, too, the first railway in the United States was constructed. True, it was built of wood, and was called a tram-way. But a car was run upon it to transport goods up and down the mountain The motion of the car was regulated by a windlass, and it was supported on runners instead of wheels. This was a very good arrangement for getting freight down the hill, but not so good for getting it up. But the wages of labor were low in every sense, since many of the Indians, demoralized by the use of those two most pestilent drugs, rum and tobacco, would do a day's work for a pint of the former and a plug of the latter.

The upper terminus of this portage was for many years merely an open landing-place for canoes and boats. In 1750, the French constructed a strong stockade-work on the bank of the river, above their barracks and storehouses. This they called Fort du Portage. It was burnt, in 1759, by Chabert Joncaire, who was in command of it when the British commenced the formidable and fatal campaign of that year against the French. After Fort

Niagara was surrendered to Sir William Johnson, Joncaire retired with his small garrison to the station on Chippewa Creek.

In less than two years the work was rebuilt in a much more substantial manner by Captain Joseph Schlosser, a German who served in the British army in that campaign. It had the outline of a tolerably regular fortification, with rude bastions and connecting curtains, surrounded by a somewhat formidable ditch. The interior plateau was a little elevated and surrounded by an earth embankment piled against the inner side of the palisades, over which its defenders could fire with great effect.

When the writer first saw its remains, the outlines and ditches of the work were distinct. Only some slight inequalities in the surface now indicate its site. Captain Schlosser was afterward promoted to the rank of colonel, and died in the fort. An oak slab, on which his name was cut, was standing at his grave just above the fort as late as the year 1808.

Some sixty rods below is still standing what is believed to be the first civilized chimney built in this part of the country. It is a large and most substantial stone structure, around which the French built their barracks. These were burnt by Joncaire on his retreat. A large dwelling-house was built to it by the English, which afforded shelter for many different occupants until it was burnt in 1813. Its last occupant, before it was destroyed, kept it as a tavern, which became a favorite place for festive and holiday gatherings. What hath been may be again. When the Falls shall have receded two miles, the brides and grooms

Mouth of the Chasm, and Brock's Monument.
Opposite page 20.

of that age will find their Cataract House near the site of old Fort Schlosser.

To the west of this old stone chimney stand the few surviving trees of the first apple orchard set out in this region. As early as 1796, it is described as being a "well-fenced orchard, containing 1200 trees." Not fifty are now standing.

Across the river from Lewiston is Queenston, so named in honor of Queen Charlotte. The battle which bears its name was fought on the 13th of October, 1813, between the American and British armies. The former crossed the river, made the attack, and carried the heights. The commander of the British forces, General Brock, and one of his aids, Colonel McDonald, were killed. The British were reënforced, and the American militia refusing to cross over to aid the Americans, the latter were obliged to return across the river, leaving a number of prisoners in the hands of the enemy. Some years afterward, the Colonial Parliament caused a fine monument to be erected on the heights to the memory of General Brock. It presents a conspicuous and imposing appearance from the terrace below.

Two miles and a quarter above Lewiston is the Devil's Hole, famous as the scene of a short supplementary campaign, made against the English, by the Seneca Indians, in 1763. Though doubtless instigated by French traders, it was a purely Indian enterprise, gotten up among themselves, and commanded by Farmer's Brother, one of the Seneca chiefs, who was a fighter as well as an orator. It was one of the best planned and most successfully executed military stratagems ever recorded. It was calculated upon the nicest balancing of facts and probabilities, and executed with unrivaled thoroughness and celerity.

It was known to the Indians that the English were in the habit, almost daily, of sending supply trains, under escort, from Fort Niagara to Fort Schlosser. After unloading at the latter post, they returned to the former. They knew also that there was a smaller supporting force of one or two companies at Lewiston, which could join the escort from Fort Niagara, in case of an extra valuable train, and that the whole force at both places was not large enough to furnish an escort of more than four hundred men; they knew that the narrow pass at the Devil's Hole was the best point to place the ambuscade; also that when the train went up they could see whether its escort was large or small, and so they would know whether they should concentrate their force to attack the larger escort, or divide it and attack the train and small escort first and the relieving force afterward. They conjectured that the train would have a small escort; but if it should have a large one, so much the better, as there would be a larger number in a small space for their balls to riddle. They conjectured also that, if the escort were small, the firing on the first attack would be heard by the soldiers at Lewiston, and that they would hurry to the relief of their comrades, not dreaming of danger before they should reach them.

The fatal result demonstrated the correctness of their reasoning. They made a double ambuscade: one for the train and escort, the other for the relieving force; and they destroyed them both, only three of the first escaping and eight of the latter. This event occurred on the 14th of September, 1773. John Stedman commanded the supply train. At the first fire of the Indians, seeing the fatal snare, he wheeled his horse at once, and, spurring him through a gauntlet of bullets, reached Schlosser in safety. A wounded soldier concealed himself in the bushes, and the drummer-boy lodged in a tree as he fell down the bank. Eight of the relieving force escaped to Fort Niagara to tell the story of their defeat.

Three miles above Schlosser is Cayuga Creek, near the mouth of which La Salle built the *Griffin*, a vessel of sixty tons burden, the first civilized craft that floated on the upper lakes, and the pioneer of an inland commerce of unrivaled growth and value. She reached Green Bay safely, but on her return voyage foundered with all on board in Lake Huron.

The French also built some small vessels on Navy Island. The reënforcements sent from Venango for the French, during the siege of Fort Niagara by Sir William Johnson, in 1759, were landed on this island. To the east of it there is a large deep basin, formed at the foot of the channel, between Grand and Buckhorn islands. The upper part of this channel being narrow, the basin appears like a bay. In this bay the French burnt and sunk the two vessels, as is supposed, which brought down the Venango reënforcements; hence the name "Burnt Ship Bay." The writer has seen the ribs and timbers of these vessels beneath the water, and caught many fine perch which had their haunts near them. The Niagara frontier was the theater of great activity during the War of 1812.

PART II.—GEOLOGY.

CHAPTER VI.

America the old world — Geologically recent origin of the Falls — Evidence thereof — Captain Williams's surveys for a ship canal — Former extent of Lake Michigan — Its outlet into the Illinois River — The Niagara barrier — How broken through — The birth of Niagara.

IF Professor Agassiz and Elie De Beaumont are correct in their geological reading, America is the old world rather than the new, and the northern portion of it, stretching from Lake Huron eastward to Labrador and northward toward the Arctic, was the first to be lifted into the genial light of the sun. And Professor Lyell has recourse to the vast stellar spaces for a standard by which to estimate "the interval of time which divides the human epoch from the origin of the coralline limestone over which the Niagara is precipitated at the Falls." "The Alps, the Pyrenees, the Himalayas," he continues, "have not only begun to exist as lofty mountain chains, but the solid materials of which they are composed have been slowly elaborated beneath the sea within the stupendous interval of ages here alluded to."

A little more than thirty years ago, Professor Agassiz

made a tour to the Upper Lakes with a class of students, for the purpose of giving them practical lessons in geology and other branches of natural science. The day was devoted to outdoor examinations of different localities, and in the evening was given a familiar lecture expository of the day's work. One of the places thus visited was Niagara, and it was the writer's good-fortune to be able to listen to the instructive lecture which followed the examination. Professor Agassiz concurs with other geologists in the opinion that the Falls were once at Lewiston, and one of the most interesting portions of the lecture was his animated description of the retrocession of the Falls, traced step by step back to their present position. From this oral exposition, from other high geological authorities, and from personal observation extending through a quarter of a century, the writer has derived the facts herein presented.

There can be no doubt that at a comparatively recent geological period the Falls of Niagara had no existence. It may suffice to mention two facts which are conclusive on this point. Dr. Houghton, geologist of the State of Michigan, stated in his report that the elevation of Lake Michigan above tide-water is five hundred and seventy-eight feet. That of Lake Erie, as shown by the surveys of the Erie Canal, is five hundred and sixty-eight feet, the difference of level between the two being ten feet. The fall or descent in the Niagara River from Lake Erie to Gill Creek, a few rods above the site of old Fort Schlosser, is twenty feet. Hence we learn that the surface of the water in Lake Michigan is thirty feet higher than that of the Niagara River near the mouth of Gill Creek. If, therefore, we find anywhere below the Falls a barrier drawn across this river that is more than thirty feet high, its water would thereby be set back to Lake Michigan. A moderate elevation above this thirty feet would serve as a safe shore-line for still water.

The existence of this barrier has been demonstrated. In the year 1835, by direction of the War Department, Captain W. G. Williams, of the United States Topographical Engineers, surveyed three routes for a canal around Niagara Falls. The first of these routes was run from the river nearly in a straight line to the head of Bloody Run, and thence a portion of the way over the terrace laid bare by the rapid subsidence of the water after the barrier had been broken through. The second route, commencing at the same point with the first,—the old Schlosser Storehouse, just above Gill Creek,—was run up the valley of the creek, through the ridge above Lewiston, at a slight depression in the general line of the hill, and thence to Lake Ontario by two different routes. The highest point in the ridge was found to be sixty feet above the surface of the water in the river at the starting point. Here, then, is found the requisite barrier—a dam thirty feet higher than the water in Lake Michigan, and having a base, as will be seen by reference to the map, of two and a half miles in breadth. This was its breadth at the time of the survey. But a careful observance of the topography of the banks on both sides of the river will show that it must have been originally not less than twice that breadth, and that the depressions now existing are the results of the denudation caused by the removal of the barrier.

While this barrier was unbroken, Lake Erie as extended would have covered all land that was not twenty-six feet higher than the present level of the river at old Schlosser landing, since the water there is sixteen feet below the level of Lake Erie. It is not difficult to trace this barrier on a good map. From old Fort Grey it stretches eastward a short distance past Batavia, and thence turns to the south through Wyoming into Cattaraugus County. In the latter county it forms the summit level of the Genesee Valley Canal. This summit is a swamp sixteen hundred and twenty-three feet above tide water, and the water runs from it northerly through the Genesee River into the Gulf of St. Lawrence, and southerly, through the Alleghany, into the Gulf of Mexico, while within a short distance rises Cattaraugus Creek which flows west into Lake Erie.

The gradual rise of the Niagara barrier as it extends to the east was demonstrated by the surveys of Captain Williams. By the Gill Creek line to Lewiston he found its elevation above the river, as has been stated, to be sixty feet. By the Cayuga Creek line to Pekin it was sixty-four feet, and by the Tonawanda Creek line to Lockport it was eighty-four feet, as is also shown by the surveys of the Erie Canal.

To the west the barrier extends from Brock's Monument to the ridge which bounds the westerly side of the valley of the Chippewa Creek, and thence around the head of Lake Ontario into the Simcoe Hills.

At that period all the islands in the Niagara River valley were submerged. The lower sections of the valleys of the Chippewa, Cayuga, Tonawanda, and Buffalo creeks were also submerged. The site of Buffalo was, probably, a small island, and many other similar islands were scattered over the broad expanse of water.

And this brings us to our second cardinal fact. Lake Michigan, having absorbed or spread over all the vast water-links in the great chain between Superior and Ontario, was the most stupendous body of fresh water on the globe. Its drainage was to the south, through the valleys of the Des Plaines, Kankakee, Illinois, and Mississippi rivers, into the Gulf of Mexico. The evidence of this fact is abundant. The survey of the Illinois Central Railroad shows that the surface of Lake Michigan is three hundred feet above the line of low water in the Ohio River at Cairo, where it joins the Mississippi. It also shows that the low-water line of the Kankakee, where the railroad crosses it, is eleven feet above the surface of the lake. This river, which forms the north-eastern branch of the Illinois, rises in the State of Indiana, near South Bend, two miles from the St. Joseph. From its very commencement at its head-springs it is a shallow channel in the middle of a swamp,—called on the maps the "Kankakee Pond,"—nearly a hundred miles long, and from two to five miles wide. On its north side, in Porter County, is a broad cove, with a small stream in the midst of it, which reaches up due north to within a stone's-throw of the south branch of the East Calumick River, which empties into the south-west corner of Lake Michigan.

More than thirty years ago, while traveling by stage from Logansport, Indiana, to Chicago, the writer was told by a fellow-passenger that it was not an unusual

GEOLOGY.   37

thing, on the occurrence of a strong north wind during the spring floods, to cross with boats from this branch of the East Calumick into the Kankakee Pond through this cove. We have not been able to obtain any authentic topographical survey which shows the elevation that must be overcome in order to effect this meeting of the waters.

Again: The river Des Plaines rises near the northern line of the State of Illinois, and running south parallel with the lake shore, at its junction with the Kankakee forms the Illinois. The Des Plaines is only ten miles west of Chicago. One of its eastern tributaries rises very near the head-waters of the south branch of the Chicago River, and often, when flooded by heavy rains, its waters flow over into the lake. At this point, also, the Jesuits and the early settlers were in the habit of crossing in their boats to the Des Plaines, and thence into the Illinois. The writer was informed by Colonel William A. Bird, the last Surveyor-in-Chief of the Boundary Commission, that when the party was at Mackinaw, in the spring of 1820, Mr. Ramsey Crooks, the adventurous and enterprising agent of John Jacob Astor, came up to that place from Joliet on the Illinois in one of the big canoes so generally used at that day for navigating the lakes, and that Mr. Crooks informed them that he crossed from the Des Plaines into Lake Michigan without taking his canoe out of the water. The deep cut in the Illinois and Michigan Canal, recently excavated by the city of Chicago in order to improve its sewer drainage, is quite uniform at its upper surface, and is sixteen to eighteen feet deep for a distance of twenty-six miles. The bottom of this cut is six feet below the lowest water-mark ever noted in the lake. At the point where the deep cut reaches the Des Plaines, it is ten feet lower than the bottom of the river. It is sixteen miles further down before the bottom of the cut and the river coincide with each other. Nearly the whole of this distance it is necessary to maintain a guard-bank, to protect the canal from the inundations of the river. Here we find there is a dam, only about twelve feet high, that once separated the waters of the lake from those of the Gulf of Mexico.

There were, therefore, two courses through which the waters of Lake Michigan could once have passed into the Illinois—the first through the Des Plaines, and the second from the head-springs of the East Calumick into the great north cove of the Kankakee Pond. When we consider the immense drainage which must have been discharged through these channels into the valley of the Illinois, we can well understand the gigantic proportions of that valley when compared with the stream which now flows through it. The perpendicular and water-worn sides of Starved Rock, below Ottawa, attest the magnitude of the lake-like floods which must once have dashed around them.

Having established the existence of the Niagara barrier, it remains to analyze its structure, and then to search out the agencies by which it was broken down. First, in regard to its organization. An examination of the locality reveals the fact that the portion of the ridge lying between old Fort Grey and Brock's Monument was of a peculiar character. At the former point the hard, compact clay had in it but a slight mixture of gray loam and sand. At the latter point, fine gravel was plentifully mingled with this loam. This latter mass, being quite porous, would rapidly become saturated with water, and its component parts be easily separated. The declivity of the high, hard, clay bank, down to the rock at the edge of the precipice, is abrupt on the American side, while on the opposite side the ascent toward Brock's Monument and above is gradual. This formation extends upward about one mile and a half, when the gravel and loam disappear, and the hard clay succeeds and continues upward with a gradual downward slope nearly to the Falls.

This upper drift was about twenty feet thick, and rested on a laminated stratum of the Niagara limestone. This stratum, though quite compact, and having its seams closely jointed, was not so thoroughly indurated as the lower strata of the Niagara group, and its thin plates were more easily displaced and broken up. The depression marked in the sixth mile of the profile referred to was evidently cut out by the waters of Fish Creek, after the barrier had been removed, since the land near the head-waters of this stream is higher than at the point where the line runs through the ridge. It is also noticeable that the ridge, at this point, approaches the brink of the escarpment more nearly than at any other, and the sharp declivity of its northern face is clearly shown on the profile in the accompanying map.

Within the last century there have been two, and perhaps more, large tidal waves on the Great Lakes. There have also been many severe gales, which have inundated the low lands around their shores, and attacked, with destructive effect, their higher banks. One of these gales is mentioned in another place. It came from about two points north of west, and, as noted, raised the water six feet on the rapids above the Falls. In the narrow portions of the river above, it must have elevated the water still more. Of course a much higher rise would have been produced by the force of such a gale acting upon the vastly increased surface of the larger lake.

The first serious impression upon the Niagara barrier must have been made by these two mighty forces. By them, undoubtedly, was made the first breach over its top, thus commencing that slow but sure denudation which finally reached the rock below. And by their aid even the rock itself was removed.

Here, then, is the composition and structure of our dam. It is thirty feet high, with a base two and a half miles certainly, and probably five, in width. How to break through it is the problem to be solved by the great inland sea which laves it, so that the water may flow onward and downward to the Atlantic.

Fortunately we have, all along the shores of our inland lakes, an annual demonstration of the method by which such problems are solved. A constant abrasion of their banks is produced by the action of water, frost, and ice. And these are the resistless elements which, by their persistent and powerful action during the lapse of ages, excavated a channel for the waters of the Niagara. The gradual upward slope of the rock and the thick upper drift broke the force of the huge waves that were occasionally dashed upon them. Their position could not have been more favorable to resist attack. It was a Malakoff of earth on a foundation of rock. Little by little the refluent waves carried back portions of the crumbled mass, and deposited them in the neighboring depressions. Slowly, wearily, desultorily, the erosion and desquamation went on. At last the upper drift was broken down, and its crumbled remains were swept from the rock.

Then the insidious forces of heat and cold, sun and frost became potent. The thin laminæ of limestone were loosened by the frost, broken up and disintegrated. At last a thin sheet of water was driven through the gorge by some fierce gale. The steep declivity of the counterscarp was then fatally attacked, and after a time its perpendicular face was laid bare. Thenceforth the elements had the top and one end of the rocky mass to work on, and they worked at a tremendous advantage. The breaking up and disintegration of the rock went on. It was gradually crumbled into sand, which was washed off by the rains or swept away by the winds. Finally a channel was excavated, of which the bottom was lower than the surface of the great lake above; the sparkling waters rushed in, dashed over the precipice, and Niagara was born.

As the water worked its way over the precipice gradually, so it would gradually excavate its channel to Lake Ontario, and it is not probable that any great inundation of the lower terrace could have occurred.

CHAPTER VII.

Composition of the terrace cut through — Why retrocession is possible — Three sections from Lewiston to the Falls — Devil's Hole — The Medina group — Recession long checked — The Whirlpool — The narrowest part of the river — The mirror — Depth of the water in the chasm — Former grand Fall.

THE water having laid bare the face of the mountain barrier from top to bottom, we are enabled to examine the composition of the mass through which it slowly cut its way. After removing the thin plates of the upper stratum, as we descend, according to Professor Hall, we find :

1. Niagara limestone — compact and geodiferous.
2. Soft argillo-calcareous shale.
3. Compact gray limestone.
4. Thin layers of green shale.
5. Gray and mottled sandstone, constituting with those below the Medina group.
6. Red shale and marl, with thin courses of sandstone near the top.
7. Gray quartzose sandstone.
8. Red shaly sandstone and marl.

Before reaching the Whirlpool the mass becomes, practically, resolved into numbers three, four, and five, the limestone, as a general rule, growing thicker and harder, and the shale also, as we follow up the stream.

The reason why retrocession of the Falls is possible is found in the occurrence of the shale noted above as underlying the rock. It is a species of indurated clay, harder or softer according to the pressure to which it may have been subjected. When protected from the action of the elements it retains its hardness, but when exposed to them it gradually softens and crumbles away. After a time the superstratum of rock, which is full of cracks and seams, is undermined and precipitated into the chasm below. If the stratum of shale lies at or near the bottom of the channel below the Falls, it will be measurably protected from the action of the elements. In this case retrocession will necessarily be very gradual. If above the Falls the shale projects upward from the channel below, then in proportion to the elevation and thickness of its stratum will be the case and rapidity of disintegration and retrocession. The shale furnishes, therefore, a good standard by which to determine the comparative rapidity with which the retrocession has been accomplished at different points.

From the base of the escarpment at Lewiston up the narrow bend in the channel above Devil's Hole, a distance of four and a quarter miles, the shale varies in thickness above the water, from one hundred and thirty feet at the commencement of the gorge, to one hundred and ten feet at the upper extremity of the bend. Here, although there is very little upward curve in the limestone, there is yet a decided curve upward in the Medina group, noticed above, composed mainly of a hard, red sandstone. It projects across the chasm, and also extends upward to near the neck of the Whirlpool, where it dips suddenly downward. The two strata of shale, becoming apparently united, follow its dip and also extend upward until they reach their maximum elevation near the middle of the Whirlpool. Thence the shale gradually dips again to the Railway Suspension Bridge, three-quarters of a mile above. For the remaining one and a half miles from this bridge to the present site of the Falls the dip is downward. We may then divide this reach of the Niagara River into three sections:

First. From Lewiston to the upper end of the Bend above Devil's Hole.

Second. Thence to the head of the rapid above the Railway Suspension Bridge.

Third. Thence to the present site of the Falls.

We are now prepared to consider these sections with reference to the retrocession of the fall of water. Through the first section the shale, as before noted, lying much above the water surface, and the superposed limestone being rather soft and thinner than at any point above, the retreat was probably quite uniform and comparatively rapid, about the same progress being made in each of the many centuries required to accomplish its whole length. Professor James Hall, in his able and interesting Report on the Geology of the Fourth District of the State of New York, suggests the probability of there having been three distinct Falls, one below the other, for some distance up-stream, when the retrocession first began. The average width of this section between the banks is one thousand feet. About one mile below its upper extremity is "Devil's Hole," a side-chasm cut out of the American bank of the river by a small stream called "Bloody Run," which, in heavy rains, forms a torrent. The "Hole" has been made by the detrition and washing out of the shale and the fall of the overlying rock. A short distance above, on the Canadian side, lies Foster's Glen, a singular and extensive lateral excavation left dry by the receding flood. The cliff at its upper end is bare and water-worn, showing that the arc or curve of the Falls must have been greater here than at any point below.

Near the upper end of this section there is a rocky cape, which juts out from the Canadian bank, and reaches nearly two-thirds of the distance across the chasm. At this point the great Fall met with a more obstinate and longer continued resistance than at any other, for the reason that the fine, firm sandstone belonging to the Medina group, as has been stated, here projects across the channel of the river, and, forming a part of its bed, rises upward several feet above the surface of the water. And here this hard, compact rock held the cataract for many centuries. The crooked channel which incessant friction and hammering finally cut through that rock is the narrowest in the river, being only two hundred and ninety-two feet wide, and the fierce rush of the water through the narrow, rock-ribbed gorge is almost appalling to the beholder. The average width between the banks of this section is about nine hundred feet.

In the second section is found the Whirlpool, one of the most interesting and attractive portions of the river.

The large basin in which it lies was cut out much more rapidly than any other part of the chasm. And this for the reason that, in addition to the thick stratum of shale, there was, underlying the channel, a large pocket, and probably, also, a broad seam or cleavage, filled with gravel and pebbles. Indeed, there is a broad and very ancient cleavage in the rock-wall on the Canadian side, extending from near the top of the bank to an unknown depth below. Its course can be traced from the north side of the pool some distance in a north-westerly direction. Of course the resistless power of the falling water was not long restrained by these feeble barriers, and here the broadest and deepest notch of any given century was made. The name, Whirlpool, is not quite accurate, since the body of water to which it is applied is rather a large eddy, in which small whirlpools are constantly forming and breaking. The spectator cannot realize the tremendous power exerted by these pools, unless there is some object floating upon the surface by which it may be demonstrated. Logs from broken rafts are frequently carried over the Falls, and, when they reach this eddy, tree-trunks from two to three feet in diameter and fifty feet long, after a few preliminary and stately gyrations, are drawn down endwise, submerged for awhile and then ejected with great force, to resume again their devious way in the resistless current. And they will often be kept in this monotonous round from four to six weeks before escaping to the rapids below.

The cleft in the bed-rock which forms the outlet of the basin is one of the narrowest parts of the river, being only four hundred feet in width. Standing on one side of this gorge, and considering that the whole volume of the water in the river is rushing through it, the spectator witnesses a manifestation of physical force which makes a more vivid impression upon his mind than even the great Fall itself. No extravagant attempt at fine writing, no studied and elaborate description, can exaggerate the wonderful beauty and fascination of this pool. It is separated from the habitations of men, at a distance from any highway, and lies secluded in the midst of a small tract of wood which has fortunately been preserved around it, in which the dark and pale greens of stately pines and cedars predominate. Within the basin the waters are rushing onward, plunging downward, leaping upward, combing over at the top in beautiful waves and ruffles of dazzling whiteness, shaded down through all the opalescent tints to the deep emerald at their base. It is ever varying, never presenting the same aspect in any two consecutive moments, and the beholder is lost in admiration as he comprehends more and more the many-sided and varied beauties of the matchless scene. No one visiting the Whirlpool should fail to go down the bank to the water's edge. On a bright summer morning, after a night shower has laid the dust, cleansed and brightened the foliage of shrub and tree, purified and glorified the atmosphere, there are few more inviting and charming views.

The remaining portion of this section is the Whirlpool rapid, a beautiful curve, reaching up just above the Railway Suspension Bridge. It was the most tumultuous and dangerous portion of the voyage once made by the *Maid of the Mist.* The water is in a perpetual tumult, a perfect embodiment of the spirit of unrest. Owing to the rapidity of the descent and the narrowness of the curve, the water is forced into a broken ridge in the center of the channel. There, in its wild tumult, it is tossed up into fanciful cones and mounds, which are crowned with a flashing coronal of liquid gems by the isolated drops and delicate spray thrown off from the whirling mass, and rising sometimes to the height of thirty feet. Standing on the bridge and looking down-stream, the spectator will see near by, on the American shore, a very good illustration of the manner in which the shale, there cropping out above the surface of the water, is worn away, leaving the superposed rock projecting beyond it.

In the third and last section the shale continues its downward dip, and at several places entirely disappears. The rock lying upon it is quite compact, and some of it very hard. The deep water into which the falling water was formerly received partially protected the shale, so that many centuries must have elapsed before the excavation of this section was completed. Its average width is eleven hundred feet.

Sixty rods below the American Fall is the upper Suspension Bridge. From this bridge, looking downward, no one can fail to be impressed with the serene and quiet beauty of the mirror below, reflecting from the surface of its emerald and apparently unfathomable depths life-size and life-like images of surrounding objects. The calm, majestic, unbroken current is in striking contrast with the fall and foam and chopping sea above.

The greatest depth of the water in mid-channel between the two Suspension Bridges, as ascertained by measuring, is two hundred feet. But it must be borne in mind that this is the depth of the water flowing above the immense mass of rock, stones, and gravel which has fallen into the channel. The bottom of the chasm, therefore, must be more than a hundred feet lower, since the fallen rocks, having tumbled down promiscuously, must occupy much more space than they did in their original bed. There are isolated points, as at the Whirlpool and Devil's Hole, where the river is wider than in any part of this section, but the depth is less. Taking into consideration both depth and width, this is the finest part of the chasm. And for this reason chiefly, when the great cataract was at a point about one hundred rods below the upper bridge, it must have presented its sublimest aspect. The secondary bank on each side of the river is here high and firm, whereby the whole mass of water must have been concentrated into a single channel of greater depth at the top of the Fall than it could have had at any other point. And here the mighty column exerted its most terrific force, rolling over the precipice in one broad, vertical curve, water falling into water, and lifting up, perpetually, that snowy veil of mist and spray which constitutes at any point its crowning beauty.

CHAPTER VIII.

Recession above the present position of the Falls — The Falls will be higher as they recede — Reason why — Professor Tyndall's prediction — Present and former accumulations of rock — Terrific power of the elements — Ice and ice bridges — Remarkable geognosy of the lake region.

THERE is probably little foundation for the apprehension which has been expressed that the recession of the chasm will ultimately reach Lake Erie and lower its level, or that the bed of the river will be worn into an inclined plane by gradual detrition, thus changing the perpendicular Fall into a tumultuous rapid. And for these reasons: The contour or arc of the Fall in its present location is much greater than it could have been at any point below. Consequently a much smaller body of water, less effective in force, is passed over any given portion of the precipice, the current being also divided by Goat and Luna islands. Also, the river bed increases in width above the Fall until it reaches Grand Island, which, being twelve miles in length by eight in width, divides the river into two broad channels, thus still further diminishing the weight and force of the falling water. The average width of the channel from Lewiston upward is one thousand feet. The present curve formed by the Falls and islands is four thousand two hundred feet. Of course the water concentrated in mass and force below the present Falls must have proved vastly more effective in disintegrating and breaking down the shale and limestone than it possibly can be at any point above. After receding half a mile further the curve will be more than a mile in extent, and hold this length for two additional miles, provided the water shall cover the bed-rock from shore to shore.

In reference to this recession, Professor Tyndall, in the closing paragraph of a lecture on Niagara, delivered before the Royal Institute, after his return to England, says: "In conclusion, we may say a word regarding the proximate future of Niagara. At the rate of excavation assigned to it by Sir Charles Lyell, namely, a foot a year, five thousand years will carry the Horseshoe Fall far higher than Goat Island. As the gorge recedes * * * it will totally drain the American branch of the river, the channel of which will in due time become cultivatable land. * * * To those who visit Niagara five millenniums hence, I leave the verification of this prediction." In his "Travels in the United States," in 1841–2, vol. I, page 27, Sir Charles Lyell says: "Mr. Bakewell calculated that, in the forty years preceding 1830, the Niagara had been going back at the rate of about a yard annually, but I conceive that one foot per year would be a more probable conjecture."

Thus it appears that the rate suggested was the result of a conjecture founded on a guess. From certain oral and written statements which we have been able to collect, we have made an estimate of the time which was required to excavate the present chasm-channel from Lewiston upward. During the last hundred and seventy-five years certain masses of rock have been known to fall from the water-covered surface of the cataract, and a statement as to the surface-measure of each mass was made. In using these data it is supposed that each break extended to the bottom of the precipice, although the whole mass did not fall at once. Of course, the substructure must have worn out before the superstructure could have gone down. Father Hennepin says that the projection of the rock on the American side was so great that "four coaches" could "drive abreast" beneath it. Seven years later, Baron La Hontan, referring to the Canadian side, says "three men" could "cross in abreast." We cannot assign less than twenty-four feet to the four coaches moving abreast. The projection on the Canadian side has diminished but little, whereas the overhang on the American side has almost entirely fallen, as is abundantly shown by the huge pile of large bowlders now lying at the foot of the precipice. Authentic accounts of similar abrasions are the following: In 1818, a mass one hundred and sixty feet long by sixty feet wide; and later in the same year a huge mass, the top surface of which was estimated at half an acre. If this estimate was correct, it would show an abrasion equivalent to nearly one foot of the whole surface of the Canadian Fall. In 1829 two other masses, equal to the first that fell in 1818, went down. In 1850 there fell a smaller mass, about fifty feet long and ten feet wide. In 1852, a triangular mass fell, which was about six hundred feet long, extending south from Goat Island beyond the Terrapin Tower, and having an average width of twenty feet. Here we have approximate data on which to base our calculations. In addition to these, it is supposed that there have been unobserved abrasions by piecemeal that equaled all the others. Combining these minor masses into one grand mass and omitting fractions, the result is a bowlder containing something more than twelve million cubic feet of rock. If this were spread over a surface one thousand feet wide and one hundred and sixty feet deep—about the average width and depth of the Falls below the ferry—it would make a block about seventy-eight feet thick. This, for one hundred and seventy-five years, is a little over five inches a year. At this rate, to cut back six miles—the present length of the chasm—would require nearly sixty thousand years, or ten thousand years for a single mile, a mere shadow of time compared with the age of the coralline limestone over which the water flows. So, if this estimate is reasonably correct, two millenniums will be exhausted before Professor Tyndall's prophecy can be fulfilled.

As to the "entire drainage of the American branch" of the river, we must be incredulous when we consider the fact that the bottom of that branch, two and a half miles above the Falls, is thirty-two feet higher than the upper surface of the water where it goes over the cliff, and that there is a continuous channel the whole distance varying from twelve to twenty feet in depth; and the further fact that, in the great syncope of the water which occurred in

1848, the topography, so to speak, of the river bottom was clearly revealed. It showed that the water was so divided, half a mile above the rapids, as to form a huge Y, through both branches of which it flowed over the precipice below, thus showing that nothing but an entire stoppage of the water can leave the American channel dry. But even if this part of Professor Tyndall's prediction should be verified, it is to be feared that his "vision" of "cultivatable land" in the case supposed will prove to be visionary. "To complete my knowledge," says Professor Tyndall, "it was necessary to see the Fall from the river below it, and long negotiations were necessary to secure the means of doing so. The only boat fit for the undertaking had been laid up for the winter, but this difficulty * * * was overcome." Two oarsmen were obtained. The elder assumed command, and "hugged" the cross-freshets instead of striking out into the smoother water. I asked him why he did so; he replied that they were directed outward and not downward." If Professor Tyndall had been at Niagara during the summer season, he would have had the opportunity, daily, of seeing the Fall "from below," and of going up or down the river on any day in a boat. All the boats (four) at the ferry are "fit for the undertaking," and all of them are, very properly, "laid up in the winter," since they would be crushed by the ice if left in the water. The oarsmen do not consider themselves very shrewd because they have discovered that it is easier to row across a current than to row against it. The party had an exciting and, according to Professor Tyndall's

Opposite page 54.   Niagara Falls, from Below.

account, a perilous trip. It is an exciting trip to a stranger, but the writer has made it so frequently that it has ceased to be a novelty.

"We reached," he says, "the Cave [of the Winds] and entered it, first by a wooden way carried over the bowlders, and then along a narrow ledge to the point eaten deepest into the shale." He also speaks of the "blinding hurricane of spray hurled against" him. This last circumstance, probably, prevented him from noticing the fact that no shale is visible in the Cave of the Winds. Its wall from the top downward, some distance beneath the place where he stood, is formed entirely of the Niagara limestone. But it is checkered by many seams, and is easily abraded by the elements.

Long-continued observation of the locality enables the writer to offer still other reasons why the Fall will never dwindle down to a rapid. As has already been noticed, the course of the river above the present Falls is a little south of west, so that it flows across the trend of the bedrock. Hence, as the Falls recede there can be no diminution in their altitude resulting from the dip of this rock. On the contrary, there is a rise of fifty feet to the head of the present rapids, and a further rise of twenty feet to the level of Lake Erie. During 1871–2, the bed of the river from Buffalo to Cayuga Creek was thoroughly examined for the purpose of locating piers for railway bridges over the stream. The greatest depth at which they found the rock—just below Black Rock dam—was forty-five feet. Generally the rock was found to be only twenty to twenty-five feet below the surface of the water.

About five miles above the present Falls there is, in the bottom of the river, a shelf of rock stretching, in nearly a straight line, across the channel to Grand Island, and having, apparently, a perpendicular face about sixteen inches deep. Its presence is indicated by a short but decided curve in the surface of the water above it, the water itself varying in depth from eleven to sixteen feet. The shelf above referred to extends under Grand Island and across the Canadian channel of the river, under which, however, its face is no longer perpendicular. If the Falls were at this point, they would be fifty-five feet higher than they are now, supposing the bed-rock to be firm. Now, by excavations made during the year 1870 for the new railway from the Suspension Bridge to Buffalo, the surface rock was found to be compact and hard, much of it unusually so. As a general rule it is well known that the greater the depth at which any given kind of rock lies below the surface, and the greater the depth to which it is penetrated, the more compact and hard it will be found to be. The rock which was found to be so hard, in excavating for the railway, lies within six feet of the surface. The deepest water in the Niagara River, between the Falls and Buffalo, is twenty-five feet. At this point, then, it would seem that the shale of the Niagara group must be at such a depth that the top of it is below the surface of the water at the bottom of the present fall. Hence, being protected from the disintegrating action of the atmosphere, and the incessant chiseling of the dashing spray, it would make a firm foundation for the hard limestone which would form the perpendicular ledge over which the water would fall. Supposing the bottom of the channel below this fall to have the same declivity as that for a mile below the present fall, the then cataract would be, as has been before stated, fifty-five feet higher than the present one. If we should allow fifty feet for a soft-surface limestone, full of cleavages and seams which might be easily broken down, still the new fall would be five feet higher than the old one. But, so far as can now be discovered, there is no geological necessity, so to speak, for making any such allowance. In the new cataract the American Fall would still be the higher, and its line across the channel nearly straight. The Canadian Fall would undoubtedly present a curve, but more gradual and uniform than the present horseshoe.

But there might possibly occur one new feature in the chasm-channel of the river as the result of future recession. That would be the presence in that channel of rocky islands, similar to that which has already formed just below the American Fall. The points at which these islands would be likely to form are those where the indurated rock of either the Medina or the Niagara group lies near the surface of the water. This probably was the case at the narrow bend below the Whirlpool, before noticed, and from thence up to the outlet of the pool. After considering what must have occurred in the last case, we may form some opinion concerning the probabilities in reference to the first.

We can hardly resist the conclusion that masses of fallen rock must have accumulated below the Whirlpool as we now see them under the American Fall. But if so, where are they? The answer to this question brings us to the consideration of the most remarkable phenomenon connected with this wonderful river. To the beholder it is matter of astonishment what can have become of the great mass of earth, rock, gravel, and bowlders, large and small, which once filled the immense chasm that lies below him. He learns that the water for a mile below the Falls is two hundred feet deep, and flows over a mass of fallen rock and stone of great depth lying below it; he sees a chasm of nearly double these dimensions, more than half of which was once filled with solid rock; he beholds the large quantities which have already fallen, which are still defiant, still breasting the ceaseless hammering of the descending flood. For centuries past this process has been going on, until a chasm seven miles long, a thousand feet wide, and, including the secondary banks, more than four hundred feet deep, has been excavated, and the material which filled it entirely removed. How? By what? Frost was the agent, ice was his delver, water his carrier, and the basin of Lake Ontario his dumping-ground. Although there is little likelihood that islands similar to Goat Island have existed in the channel from Lewiston upward, still it is probable that, when the Fall receded from the rocky cape below the Whirlpool up to the pool, it left masses of rock, large and small, lying on the rocky floor and projecting above the surface of the water. As there were no islands above, there were no broken, tumultuous rapids. As has been before remarked, the water poured over in one broad, deep, resistless flood. When frozen by the intense cold of winter, the great cakes of ice would descend with crushing force on these rocks. The smaller ones would be broken, pulverized, and swept downstream, the channel for the water would be enlarged gradually, and the larger masses thus partially undermined. Then the spray and dashing water would freeze and the ice accumulate upon them until they were toppled over. Then the falling ice would recommence its chipping labors, and with every piece of ice knocked off, a portion of the rock would go with it. Finally, as the cold continued, the master force, the mightiest of mechanical powers, would be brought into action. The vast quantities of ice pouring over the precipice would freeze together, agglomerate, and form an ice-bridge. The roof being formed, the succeeding cakes of ice would be drawn under, and, raising it, be frozen to it. This process goes on. Every piece of rock above and below the surface is embraced in a relentless icy grip. Millions of tons are frozen fast together. The water and ice continue to plunge over the precipice. The principle of the hydrostatic press is made effective. Then commences a crushing and grinding process which is perfectly terrific. Under the resistless pressure brought to bear upon it, the huge mass moves half an inch in one direction, and an hundred cubic feet of rock are crushed to powder. There is a pause. Then again the immense structure moves half an inch another way, and once more the crumbling atoms attest its awful power. This goes on for weeks continuously. Finally the temperature changes. The sunlight becomes potent; the ice ceases to form; the warm rays loosen the grip of the ice-bridge along the borders of the chasm below. The water becomes more abundant; the bridge rises, bringing in its icy grasp whatever it had attached itself to beneath; it breaks up into masses of different dimensions: each mass starts downward with the growing current, breaking down or filing off everything with which it comes in contact. Fearful sounds come up from the hidden depths, from the mills which are slowly pulverizing the massive rock. The smaller bits and finer particles, after filling the interstices between the larger rocks in the bottom of the chasm, are borne lakeward. The heavier portions make a part of the journey this year; they will make another part next year, and another the next, being constantly disintegrated and pulverized.

This work has been going on for many centuries. The result is seen in the vast bar of unknown depth which is spread over the bottom of Lake Ontario around the mouth of the river. On the inner side of the bar the water is from sixty to eighty feet deep, on the bar it is twenty-five feet deep, and outside of it in the lake it reaches a depth of six hundred feet.

And finally, to the force we have been considering, more than to any other, it is probable that all the coming generations of men will be indebted for a grand and perpendicular Fall somewhere between its present location and Lake St. Clair; for it must be remembered that the bottom of Lake Erie is only fourteen feet lower than the crest of the present Fall, and the bottom of Lake St. Clair is sixty-two feet higher. It may also be considered that the corniferous limestone of the Onondaga group — which

Great Icicles under the American Fall.
Opposite page 60.

succeeds the Niagara group as we approach Lake Erie — is more competent to maintain a perpendicular face than is the limestone of the latter group.

We may here appropriately notice a remarkable feature in the geognosy of the earth's surface from Lake Huron to the Gulf of St. Lawrence. We have before stated that the elevation of that lake above tide-water is five hundred and seventy-eight feet. But its depth, according to Dr. Houghton, is one thousand feet. If this statement is correct, the bottom of it is four hundred and twenty-two feet below the sea-level. The elevation of Lake St. Clair is five hundred and seventy feet. But its depth is only twenty feet, leaving its bottom five hundred and fifty feet above the sea-level. The elevation of Lake Erie is five hundred and sixty-eight feet. But it is only eighty-four feet deep, making it four hundred and eighty-four feet above the sea-level. From Lake Erie to Lake Ontario there is a descent of three hundred and thirty-six feet. But the latter lake is six hundred feet deep, and its elevation two hundred and thirty-two feet. Hence the bottom of it is three hundred and sixty-eight feet below the sea-level. From the outlet of Lake Ontario the St. Lawrence River flows eight hundred and twenty miles to tide-water, falling two hundred and thirty-two feet in this distance. The water from the springs at the bottom of Lake Huron is compelled to climb a mountain nine hundred and eighty feet high before it can start on this long oceanward journey.

PART III.

LOCAL HISTORY AND INCIDENTS.

CHAPTER IX.

Forty years since — Niagara in winter — Frozen spray — Ice foliage and ice apples — Ice moss — Frozen fog — Ice islands — Ice statues — Sleigh-riding on the American rapids — Boys coasting on them — Ice gorges.

IF the first white man who saw Niagara could have been certain that he was the first to see it, and had simply recorded the fact with whatever note or comment, he would have secured for himself that species of immortality which accrues to such as are connected with those first and last events and things in which all men feel a certain interest. But he failed to improve his opportunity, and Father Hennepin was the first, so far as known, to profit by such neglect, and his somewhat crude and exaggerated description of the Falls has been often quoted and is well known. So long as "waters flow and trees grow" it will continue to be read by successive generations. The French missionaries and traders who followed him seem to have been too much occupied in saving souls or in seeking for gold to spend much time in contemplating the cataract, or to waste much sentiment in writing about it. And so it happens that, considering its fame, very little has been written, or rather published, concerning it.

Seventy years ago, the few travelers who were drawn to the vicinity by interest or curiosity were obliged to approach it by Indian trails, or rude corduroy roads, through dense and dark forests. Within the solitude of their deep shadows, beneath their protecting arms, was hidden one of the sublimest works of the physical creation. The scene was grand, impressive, almost oppressive, not less sublime than the Alps or the ocean, but more fascinating, more companionable, than either.

Niagara we can take to our hearts. We realize its majesty and its beauty, but we are never obliged to challenge its power. Its surroundings and accessories are calm and peaceful. Even in all the treacherous and bloody warfare of savage Indians it was neutral ground. It was a forest city of refuge for contending tribes. The generous, noble, and peaceful Niagaras—a people, according to M. Charlevoix, "larger, stronger, and better formed than any other savages," and who lived upon its borders—were called by the whites and the neighboring tribes the Neuter Nation.

The crafty Hurons, the unwarlike Eries, the invincible league formed by the six aggressive and conquering tribes composing the Iroquois confederacy,—the

Mohawks, the Oneidas, the Onondagas, the Cayugas, the Senecas, and the Tuscaroras,—all extinguished the torch, buried the tomahawk, and smoked the calumet when they came to the shores of the Niagara, and sat down within sight of its incense cloud, and listened to its perpetual anthem. In succeeding contests between the whites, on two occasions only was nature's repose here disturbed by the din of battle—first, in the running fight at Chippewa, and again at the obstinate and bloody struggle of Lundy's Lane.

During the War of 1812, in which these actions occurred, the dense forest which lay outside of the old belt of French occupation was first extensively and persistently attacked, the sunlight being let in upon comfortable log-cabins and fruitful fields. The Indian trail and corduroy "shake" were superseded by more civilized and comfortable highways. Post routes were opened and public conveyances established. For many years, however, the two principal ways of access to Niagara were by the Ridge road, from the Genessee Falls—now Rochester—and the river road on the Canadian side from Buffalo to Drummondville.

Some forty years ago, and for many years thereafter, Niagara was, emphatically, a pleasant and attractive watering-place; the town was quiet; the accommodations were comfortable; the people were kind, considerate, and attentive; guides were civil, intelligent, and truthful; conveyances were good, and were in charge of careful and respectable attendants; commissions were unknown; "scalping" was left to the Indians; nobody was annoyed or importuned; the flowers bloomed, the birds caroled, the full-leaved trees furnished refreshing shade, and the air was balmy. Then the lowing of cows in the street, the guttural note of the swine, and the voice of the solicitor were not heard. Elderly people came to stay for pleasant recreation and quiet enjoyment; younger people to "bill and coo" and dance. Now all that is changed. A contemporary orator once described the moral status of a famous stock-jobbing locality by saying that "ten thousand a year is the Sermon on the Mount for Wall street." The same gospel is popular at Niagara.

Whoso has seen Niagara only in summer has but half seen it. In winter its beauties are not diminished, while the accessories due to the season are numerous and varied. After two or three weeks of intensely cold weather many beautiful and fantastic scenes are presented around the Falls.

The different varieties of stalactites and stalagmites hanging from or apparently supporting the projecting rocks along the side walls of the deep chasm, the ice islands which grow on the bars and around the rocks in the river, the white caps and hoods which are formed on the rocks below, the fanciful statuary and statuesque forms which gather on and around the trees and bushes, are all curious and interesting. Exceedingly beautiful are the white vestments of frozen spray with which everything in the immediate vicinity is robed and shielded; and beautiful, too, are the clusters of ice apples which tip the extremities of the branches of the evergreen trees.

There is something marvelous in the purity and whiteness of congealed spray. One might think it to be frozen sunlight. And when, by reason of an angle or a curve, it is thrown into shadow, one sees where the rainbow has been caught and frozen in. After a day of sunshine which has been sufficiently warm to fill the atmosphere with aqueous vapor, if a sharp, still, cold night succeed, and if on this there break a clear, calm morning, the scene presented is one of unique and enchanting beauty.

The frozen spray on every boll, limb, and twig of tree and shrub, on every stiffened blade of grass, on every rigid stem and tendril of the vines, is covered over with a fine white powder, a frosty bloom, from which there springs a line of delicate frost-spines, forming a perfect fringe of ice-moss, than which nothing more fanciful nor more beautiful can be imagined.

Then, as the day advances, the increasing warmth of the sun's rays dissolves this fairy frost-work and spreads it like a delicate varnish over the solid spray, giving it a brilliant polish rivaling the luster of the rarest gems; the mid-morning breeze sets in motion this flashing, dazzling forest, which varies its color as the sunlight-angle varies; and finally, when the waxing warmth and growing breeze loosen the hold of the icy covering in the tree-tops, and it drops to the still solid surface in the shade beneath,—the tiny particles

Opposite page 66. Winter Foliage.

with a silver tinkle and the larger pieces with the sharp, rattling sound of the castanet,—the ear is charmed with a wild, dashing rataplan, while a scene of strange enchantment challenges the admiration of the spectator.

Even more beautiful and fairy-like, if possible, is the garment of frozen fog with which all external objects are adorned and etherealized when the spring advances and the temperature of the water is raised. As the sharp, still night wears on, the light mists begin to rise, and when the morning breaks, the river is buried in a deep, dense bank of fog. A gentle wave of air bears it landward; its progress is stayed by everything with which it comes in contact, and as soon as its motion is arrested it freezes sufficiently to adhere to whatever it touches. So it grows upon itself, and all things are soon covered half an inch in depth with a most delicate and fragile white fringe of frozen fog. The morning sun dispels the mist, and in an hour the gay frostwork vanishes.

The ice islands are sometimes extensive. In the year 1856 the whole of the rocky bar above Goat Island was covered with ice, piled together in a rough heap, the lower end of which rested on Goat Island and the three Moss Islands lying outside of it, all of which were visited by different persons passing over this new route.

The ice formed on the rocks below the American Fall, stretched upward, reached the edge of the precipice just north of the Little Horseshoe, continued up-stream above

Chapin's Island, spread out laterally from that to Goat Island on the south, and over nearly half of the American rapids to the north. At the brow of the precipice it accumulated upward until it formed a ridge some forty feet high. About fifteen rods up-stream another ridge was formed of half the height of the first. Every rock projecting upward bore an immense ice-cap. Around and between these mounds and caps horses were driven to sleighs, albeit the course was not favorable for quick time. The boys drew their sleds to the top of the large mound, slid down it, up-stream, and nearly to the top of the smaller hill.

On the lower or down-stream side, they would have had a clear course to the water below, at the brink of the Falls, and might have made "time" compared with which Dexter's minimum would have seemed only a funeral march. But with all Young America's passion for speed, he declined to try this route. The writer walked over the south end of Luna Island, above the tops of the trees.

The ice-bridge of that year filled the whole chasm from the Railway Suspension Bridge up past the American Fall. When the ice broke up in the spring, such immense quantities were carried down that a strong northerly wind across Lake Ontario caused an ice-jam at Fort Niagara. The ice accumulated and set back until it reached the Whirlpool, and could be crossed at any point between the Whirlpool and the Fort. It was lifted up about sixty feet above the surface, and spread out over both shores, crushing and destroying everything with which it came in

Opposite page 69.   Ice Bridge and Frost Freaks.

contact. Many persons from different parts of the country visited the extraordinary scene.

At Lewiston the writer, with many others, saw a most remarkable illustration of the terrific power of this hydrostatic press. Just below the village, on the American side, there stood, about two rods from high-water mark, a sound, thrifty, tough white-oak tree, perhaps a hundred years old, and two feet in diameter. The ice, moved by the water, struck it near the ground and pressed it outward and upward, until it was actually pulled up by the roots —or rather some of the roots were broken and others were pulled out—and landed twenty feet farther away from the chasm.

Those who watched the operation stated that, from the time the ice touched the tree until it was landed on the bank above, the motion of the ice could not be detected by the eye.

Slowly, steadily, surely it pressed on. Suddenly there would be an explosion, sharp and loud, when a root gave way. No motion in the ice or tree could be discovered. After a lapse of two or three hours another sharp crack would give notice of another fracture. Thus the ice pressed gradually on, and in ten hours the work was done. A thousandth part of this force would pulverize a bowlder of adamant. We need not wonder, therefore, that the river Niagara keeps its channel clear.

In the ice-gorge of 1866 the ice was set back to the upper end of the Whirlpool, over which it was twenty feet deep. The Whirlpool rapid was subdued nearly to an unbroken current, which all the way below to Lake

Ontario was reduced to a gentle flow of quiet waters. Never was there a sublimer contest of the great forces of nature. The frost laid its hand upon the raging torrent and it was still.

The winter of 1875 was intensely cold. The singular figures represented in the illustrations—the eagle, dog, baboon, and others—are exact reproductions of the real chance-work of the frost of that season. The long-continued prevalence of the south-west wind fastened to every object facing it a border or apron of dazzling whiteness, and more than five feet thick. The ice mountain below the American Fall, reaching nearly to the top of the precipice, was appropriated as a "coasting" course, and furnished most exhilarating sport to the people who used it. A large number of visitors came from all directions, and, on the 22d of February, fifteen hundred were assembled to see the extraordinary exhibition.

In the coldest winters, the ice-bridges cannot be less than two hundred and fifty feet thick. The ice-bridge of 1875 formed on the 6th and 7th of May, was crossed on the 8th, and broke up on the 14th—the only one ever known in the river so late in the spring.

Coasting below the American Fall.

CHAPTER X.

Judge Porter — General Porter — Goat Island — Origin of its name — Early dates found cut in the bark of trees and in the rock — Professor Kalm's wonderful story — Bridges to the Island — Method of construction — Red Jacket — Anecdotes — Grand Island — Major Noah and the New Jerusalem — The Stone Tower — The Biddle Stairs — Sam Patch — Depth of water on the Horseshoe — Ships sent over the Falls.

IN preparing this narrative, the writer has had the good fortune to listen to many recitals of facts and incidents by the late Judge Augustus Porter and the late General Peter B. Porter, whose names are intimately and honorably connected with the more recent history, not only of this particular locality but of the Empire State.

Judge Porter, after having spent several years in surveying and lotting large portions of the territory of Western New York and the Western Reserve in Ohio, came from Canandaigua to Niagara Falls with his family in June, 1806, where he continued to live until his death, nearly fifty years afterward.

General Porter settled as a lawyer at Canandaigua in 1795, removed to Black Rock in 1810, and to Niagara Falls in 1838.

In 1805, the two brothers became interested with others in the purchase from the State of New York of four lots in the Mile Strip lying both above and below the Falls.

A few years later, they purchased not only the interest of their partners in these lots, but other lands at different points along this strip. In 1814, they bought of Samuel Sherwood a paper since named a *float*—an instrument given by the State authorizing the bearer to locate two hundred acres of any of the unsold or unappropriated lands belonging to the State. This float they fortunately anchored on Goat Island and the islands adjacent thereto lying "immediately above and adjoining the Great Falls."

The origin of the name of Goat Island is as follows: Mr. John Stedman, who came into the country in 1760, had cleared a portion of the upper end of the island, and in the summer of 1779 he placed on it an aged and dignified male goat. The following winter was very severe, navigation to the island was impracticable, and the goat fell a victim to the intense cold. Since which the scene of his exile and death has been called Goat Island.

By the terms of the Treaty of Ghent, December 24, 1814, the boundary line between Great Britain and the United States, on the Niagara frontier, was to run through the deepest water along the river-courses and through the center of the Great Lakes. As the deepest water, at this point, is in the center of the Horseshoe Fall, the islands in the river fell to the Americans. General Porter, acting as Commissioner for the United States, proposed to call the largest one Iris Island, and it was so printed on the boundary maps. But the public adhered to the old name of Goat Island.

One of the early chronicles states that the island contained two hundred and fifty acres of land. At the present time there are in it less than seventy. A strip some ten rods wide by eighty rods long has been worn away from the southern side of it since 1818, when Judge Porter made the first road around it.

The earliest date he found on the island was 1765, carved on a beech-tree. The earliest date cut in the rock on the main-land was 1645. Human bones and arrowheads were found on the island. The Indians went to it with their canoes, which they paddled up and down in the comparatively quiet water lying on the rocky bar which extends upward nearly a mile above the head of the island.

Notwithstanding this fact, the Swedish naturalist, Kalm, who visited the place in 1750, relates a fabulous story of two Indians who, on a hunting excursion above the Falls, drank too freely from "two bottles of French brandy" which they brought from Fort Niagara; becoming drowsy, they laid themselves down in the bottom of their canoe for a nap.

The canoe swung off shore and floated down-stream. Nearing the rapids, the noise awakened one of them, who had apparently been more fortunate in learning the English language from the French than most of his tribe, for, seeing their perilous situation, he exclaimed: "We are gone!" But the two plied their paddles with such aboriginal vigor that they succeeded in landing on

Goat Island. From the sequel it would seem that they must have destroyed or lost their canoe. Finding no houses of refreshment, nor cairns of stores left by former explorers, and most naturally getting hungry, they concluded it would be desirable to get back to the fort—a wish more easily expressed than accomplished.

But it was necessary for them to "do or die." So, as the story runs, they stripped the bark from the basswood trees, and with it made a ladder long enough to reach from a tree standing on the edge of the precipice at the foot of the island down to the water below.

After dropping their ladder they followed it downward. Reaching the water, and being good swimmers, they plunged in with great glee, expecting to be able to swim across to the opposite shore, which they could easily climb. But the counter current forced them back to the island.

After being a good deal bruised on the rocks, they were compelled to abandon the attempt to cross, and then returned up their ladder to the island. There, after much whooping, they attracted the notice of other Indians on the shore. These reported the situation at the fort, and the commandant sent up a party of whites and Indians to rescue them. They brought with them four light pike-poles. Going to a point opposite the head of the island, they exchanged salutations with the new Crusoes, and began preparations for their rescue. Two Indians volunteered to undertake the task. "They took leave of all their friends as if they were going to their death." Each Indian rescuer, according to the wondrous fable, took two pike-poles and *waded* across the channel to the island, gave each of the Crusoes a pike-pole, and then the four waded back to the main-land, where they were joyfully received by their anxious, waiting friends, after having been " nine days on the island."

Remembering that the water in mid-channel is twelve feet deep, with a twelve-mile current, we must concede this to be the most marvelous of all aquatic achievements.

In 1817 Judge Porter built the first bridge to Goat Island, about forty rods above the present bridge. In the following spring the large cakes of ice from the river above, not being sufficiently broken up by the short stretch of rapids over which they passed, struck the bridge with terrific force, and carried away the greater part of it. With the courage and enterprise of a New-Englander, the next season he constructed another bridge farther down, on the present site, rightly judging that the ice would be so much broken up before reaching it as to be harmless.

That bridge, with constant repairs and one almost entire renewal, stood firm in its place until the year 1856, when it was removed to make room for the present iron bridge. The old piers were much enlarged and strengthened, and also raised about three feet higher to receive the new bridge. As nearly every stranger inquires how the first bridge was carried over the turbulent waters, a brief description of the process may be acceptable. First, a strong bulkhead was built in the shallow water next to the shore; a solid backing was put in behind this, and the upper surface properly graded and well floored with plank. Strong rollers were placed parallel with the stream and fastened to the floor. In the old forest then standing near by were many noble oaks, of different sizes and great length. A number of these were felled and hewed "tapering," as it was termed, so that, when finished, they were about eighteen inches square at the butt, fifteen at the top, and eighty feet long. Through the small ends were bored large auger-holes. These sticks were placed, as required, on the rollers, at right angles to the stream, the small ends over the water, and the shore ends heavily weighted down.

The first stick being properly placed, levers were applied to the rollers and the stick was run out until the small end reached an eddy in the water. Then another similar stick was run out in like manner, parallel to the first, and about six feet from it. A few light, strong planks were placed across and made fast. Two men were provided each with strong, iron-pointed pike-staffs, each staff having in its upper end a hole, through which was drawn some ten feet of new rope. Thus provided, they walked out on the timbers, drove their iron pikes down among the stones, and tied them fast to the timbers. Thus the whole problem was solved. Around these pike-staffs the first pier was built and filled with stone. Then other timbers were run out, all were planked over, and the first span was completed. The other spans were laid in the same way.

The great Indian chief and orator, Red Jacket, occasionally visited Judge and General Porter—the latter

Second Moss-Island Bridge.

Opposite page 76.

then living at Black Rock. Judge Porter told this anecdote of the chief: He visited the Falls while the mechanics were stretching the timbers across the rapids for the second bridge. He sat for a long time on a pile of plank, watching their operations. His mind seemed to be busy both with the past and the present, reflecting upon the vast territory his race once possessed, and intensely conscious of the fact that it was theirs no longer. Apparently mortified, and vexed that its paleface owners should so successfully develop and improve it, he rose from his seat, and, uttering the well-known Indian guttural "Ugh, ugh!" he exclaimed: "D——n Yankee! d——n Yankee!" Then, gathering his blanket-cloak around him, with his usual dignity and downcast eyes, he slowly walked away, and never returned to the spot.

Before parting with the distinguished chief, we will repeat after General Porter two other anecdotes characteristic of him. He lived not far from Buffalo, on the Seneca Reservation, and frequently visited the late General Wadsworth, at Geneseo. Indeed, his visits grew to be somewhat perplexing, for the great chief must be entertained personally by the host of the establishment.

Of course he was a "teetotaler"—only in one way. When he got a glass of good liquor he drank the whole of it. He was very fond of the rich apple-juice of the Geneseo orchards. Having repeated his visits to General Wadsworth, at one time, with rather inconvenient frequency, and coming one day when the General saw that he had been drinking pretty freely somewhere else, his host concluded he would not offer him the usual refreshments. In due time, therefore, Red Jacket rose and excused himself. As he was leaving the room the orator said, "General, hear!" "Well, what, Red Jacket?" To which he replied with great gravity: "General, when I get home to my people, and they ask me how your cider tasted, what shall I tell them?" Of course he got the cider.

His determined and constant opposition to the sale of the lands belonging to the Indians is well known. At the council held at Buffalo Creek, in 1811, he was selected by the Indians to answer the proposition of a New York land company to buy more land. The Indians refused to sell, although, as usual, the company only wanted "a small tract." To illustrate the system, after the speech-making was over, Red Jacket placed half a dozen Indians on a log, which lay near by. They did not sit very close together, but had plenty of room. He then took a white man who wanted "a small tract," and making the Indians at one end "move up," he put the white man beside them. Then he brought another "small-tract" white man, and making the aborigines "move up" once more, the Indian on the end was obliged to rise from the log. He repeated this process until but one of the original occupants was left on the log. Then suddenly he shoved him off, put a white man in his place, and turning to the land agent said: "See what one *small tract* means; white man *all*, Indian *nothing*."

Colonel William L. Stone, in his "Life of Red Jacket," relates the following: In 1816, after Red Jacket took up his residence on Buffalo Creek, east of the city, a young French count traveling through the country made a brief stay at Buffalo, whence he sent a request to the sachem to visit him at his hotel.

Red Jacket, in reply, informed the young nobleman that if he wished to see the old chief he would give him a welcome greeting at his cabin. The count sent again to say that he was much fatigued by his journey of four thousand miles, which he had made for the purpose of seeing the celebrated Indian orator, Red Jacket, and thought it strange that he should not be willing to come four miles to meet him. But the proud and shrewd old chief replied that he thought it still more strange, after the count had traveled so great a distance for that purpose, that he should halt only a few miles from the home of the man he had come so far to see. The count finally visited the sachem at his house, and was much pleased with the dignity and wisdom of his savage host. The point of etiquette having been satisfactorily settled, the chief accepted an invitation to dinner, and was no doubt able to tell his people how the count's "cider" tasted.

In 1819, when the boundary commissioners ran the line through the Niagara River, Grand Island fell to the United States, under the rule that that line should be in the center of the main channel. To ascertain this, accurate measurements were made, by which it was found that 12,802,750 cubic feet of water passed through the Canadian channel, and 8,540,080 through the American channel. To test the accuracy of these measurements, the quantity passing in the narrow channel at Black Rock was determined by the same method, and was found to be 21,549,590 cubic feet, thus substantially corroborating the first two measurements.

The Indian name of Grand Island is Owanunga. In 1825, Mr. M. M. Noah, a politician of the last generation, took some preliminary steps for reëstablishing the lost nationality of the Jews upon this island, where a New Jerusalem was to be founded. Assuming the title of "Judge of Israel," he appeared at Buffalo in September for the purpose of founding the new nation and city. A meeting was held in old St. Paul's Church, at which, with the aid of a militia company, martial music, and masonic rites, the remarkable initiatory proceedings took place.

The self-constituted judge presented himself arrayed in gorgeous robes of office, consisting of a rich black cloth tunic, covered by a capacious mantle of crimson silk trimmed with ermine, and having a richly embossed golden medal hanging from his neck. After what, in the account published in his own paper of the day's proceedings, he called "impressive and unique ceremonies," he read a proclamation to "all the Jews throughout the world," informing them "that an Asylum was prepared and offered to them," and that he did "revive, renew, and establish (in the Lord's name), the government of the Jewish nation, * * * confirming and perpetuating all our rights and privileges, our rank and power, among the nations of the earth as they existed and were recognized under the government of the Judges." He also ordered a census to be taken of all the Hebrews in the world, and levied a capitation tax of three shekels—about one dollar and sixty cents—" to pay the expenses of re-organizing the government and assisting emigrants." He had prepared a " foundation stone," which was afterward erected on the site of the new city, and which bore the following inscription:

" Hear, O Israel, the Lord
is our God—the Lord is one."

"ARARAT,
A CITY OF REFUGE FOR THE JEWS,
FOUNDED BY MORDECAI MANUEL NOAH,
IN THE MONTH OF TISRI 5586—SEPT. 1825,
IN THE FIFTIETH YEAR OF
AMERICAN INDEPENDENCE."

After the meeting at St. Paul's, the "Judge" returned at once to New York, and, like the great early ruler of his nation, he only saw the land of promise, as he never crossed to the island.

The strong round tower, called the Terrapin Tower, which stood near Goat Island, not far from the precipice, was built in 1833, of stones gathered in the vicinity. It was forty-five feet high, and twelve feet in diameter at the base. So much was said in 1873 about the growing insecurity of the tower that it was taken down.

The Biddle Staircase was named for Mr. Nicholas Biddle, of Philadelphia, who contributed a sum of money toward its construction. It was erected in 1829. The shaft is eighty feet high and firmly fastened to the rock. The stairs are spiral, winding round it from top to bottom. Near the foot of these stairs, at the water's edge, Samuel Patch, who wished to demonstrate to the world that "some things could be done as well as others," set up a ladder one hundred feet high, from which he made two leaps into the water below. Going thence to Rochester, he took another leap near the Genesee Falls, which proved to be his last.

The depth of water on the Horseshoe Fall is a subject of speculation with every visitor. It was correctly determined in 1827. In the autumn of that year, the ship *Michigan*, having been condemned as unseaworthy, was purchased by a few persons, and sent over the Falls. Her hull was eighteen feet deep. It filled going down the rapids, and went over the Horseshoe Fall with some water above the deck, indicating that there must have been at least twenty feet of water above the rock. This voyage of the *Michigan* was an event of the day. A glowing hand-bill, charged with bold type and sensational tropes, announced that "The Pirate *Michigan*, with a cargo of furious animals," would "pass the great rapids and the Falls of Niagara," on the "eighth of September, 1827." She would sail "through the white-tossing and deep-rolling rapids of Niagara, and down its grand precipice into the basin below." Entertainment was promised "for all who may visit the Falls on the present occasion, which will, for its novelty and the remarkable spectacle it will present, be unequaled in the annals of *infernal* navigation." Considering that the Falls could be reached only by road conveyances, the gathering of people was very large. The voyage was successfully made, and the "cargo of live animals" duly deposited in the "basin below," except a bear which left the ship near the center of the rapids and swam ashore, but was recaptured.

Two enterprising individuals made arrangements to supply the people assembled on the island with refreshments. They had an ample spread of tables and an abundant supply of provisions. As there was much delay in getting the vessel down the river, the people got impatient and hungry. They took their places at the tables. When their appetites were nearly satisfied, notice was given that the ship was coming, whereupon they departed hurriedly, forgetting to leave the equivalent half-dollar for the benefit of the purveyors.

In after years, one of the proprietors of this unexpected "free lunch"—the late General Whitney—established here one of the best hotels in the country, and left his heirs an ample fortune.

A few geese in the cargo were only badly confused by their unusual plunge, and were afterward picked up from boats. It was noticed as being a little singular that geese which went over the Falls in the Pirate *Michigan* were for sale at extravagant prices all the next season.

Another condemned vessel of about five hundred tons burden, the *Detroit*, which had belonged to Commodore Perry's victorious fleet, was sent down the rapids in 1841. A large concourse of people assembled from all parts of the country to witness the spectacle. Her rolling and plunging in the rapids were fearful, until about midway of them she stuck fast on a bar, where she lay until knocked to pieces by the ice. From Baron La Hontan we know that the Indians went on the water, just below the Falls, in their canoes, to gather the game which had been swept over them. For more than a hundred years there has been a ferry of skiff and yawl boats at this point, and in all that time not one serious accident has happened.

CHAPTER XI.

Joel R. Robinson, the first and last navigator of the Rapids — Rescue of Chapin — Rescue of Allen — He takes the *Maid of the Mist* through the Whirlpool — His companions — Effect upon Robinson — Biographical notice — His grave unmarked.

THE history of the navigation of the Rapids of Niagara may be appropriately concluded in this chapter, which is devoted to a notice of the remarkable man who began it, who had no rival and has left no successor in it — Joel R. Robinson.

In the summer of 1838, while some extensive repairs were being made on the main bridge to Goat Island, a mechanic named Chapin fell from the lower side of it into the rapids, about ten rods from the Bath Island shore. The swift current bore him toward the first small island lying below the bridge. Knowing how to swim, he made a desperate and successful effort to reach it. It is hardly more than thirty feet square, and is covered with cedars and hemlocks. Saved from drowning, he seemed likely to fall a victim to starvation. All thoughts were then turned to Robinson, and not in vain. He launched his light red skiff from the foot of Bath Island, picked his way cautiously and skillfully through the rapids to the little island, took Chapin in and brought him safely to the shore, much to the relief of the spectators, who gave expression to their appreciation of Robinson's service by a moderate contribution.

In the summer of 1841, a Mr. Allen started for Chippewa in a boat just before sunset. Being anxious to get across before dark, he plied his oars with such vigor that one of them broke when he was about opposite the middle Sister. With the remaining oar he tried to make the head of Goat Island. The current, however, set too strongly toward the great Canadian Rapids, and his only hope was to reach the outer Sister. Nearing this, and not being able to run his boat upon it, he sprang out, and, being a good swimmer, by a vigorous effort succeeded in getting ashore. Certain of having a lonely if not an unpleasant night, and being the fortunate possessor of two stray matches, he lighted a fire and solaced himself with his thoughts and his pipe. Next morning, taking off his red flannel shirt, he raised a signal of distress. Toward noon the unusual smoke and the red flag attracted attention. The situation was soon ascertained, and Robinson informed of it. Not long after noon, the little red skiff was carried across Goat Island and launched in the channel just below the Moss Islands. Robinson then pulled himself across to the foot of the middle Sister, and tried in vain to find a point where he could cross to the outer one. Approaching darkness compelled him to suspend operations. He rowed back to Goat Island, got some refreshments, returned to the middle Sister, threw the food across to Allen, and then left him to his second night of solitude. The next day

Opposite page 96.   Joel R. Robinson.

Robinson took with him two long, light, strong cords, with a properly shaped piece of lead weighing about a pound. Tying the lead to one of the cords he threw it across to Allen. Robinson fastened the other end of Allen's cord to the bow of the skiff; then attaching his own cord to the skiff also, he shoved it off. Allen drew it to himself, got into it, pushed off, and Robinson drew him to where he stood on the middle island. Then seating Allen in the stern of the skiff he returned across the rapids to Goat Island, where both were assisted up the bank by the spectators, and the little craft, too, which seemed to be almost as much an object of curiosity with the crowd as Robinson himself.

This was the second person rescued by Robinson from islands which had been considered wholly inaccessible. It is no exaggeration to say that there was not another man in the country who could have saved Chapin and Allen as he did.

In the summer of 1855 a canal-boat, with two men and a dog in it, was discovered in the strong current near Grass Island. The men, finding they could not save the large boat, took to their small one and got ashore, leaving the dog to his fate. The abandoned craft floated down and lodged on the rocks on the south side of Goat Island, and about twenty rods above the ledge over which the rapids make the first perpendicular break. There were left in the boat a watch, a gun, and some articles of clothing. The owner offered Robinson a liberal salvage if he would recover the property. Taking one of his sons with him, he started the little red skiff from the head of the hydraulic canal, half a mile above the island, shot across the American channel, and ran directly to the boat. Holding the skiff to it himself, the young man got on board and secured the valuables. The dog had escaped during the night. Leaving the canal-boat, Robinson ran down the ledge between the second and third Moss Islands, and thence to Goat Island. On going over the ledge he had occasion to exercise that quickness of apprehension and presence of mind for which he was so noted. The water was rather lower than he had calculated, and on reaching the top of the ledge the bottom of the skiff near the bow struck the rock. Instantly he sprang to the stern, freed the skiff, and made the descent safely. If the stern had swung athwart the current, the skiff would certainly have been wrecked.

In the year 1846, a small steamer was built in the eddy just above the Railway Suspension Bridge, to run up to the Falls. She was very appropriately named *The Maid of the Mist*. Her engine was rather weak, but she safely accomplished the trip. As, however, she took passengers aboard only from the Canadian side, she could pay little more than expenses. In 1854 a larger, better boat, with a more powerful engine, the new *Maid of the Mist*, was put on the route, and as she took passengers from both sides of the river, many thousands of persons made the exciting and impressive voyage up to the Falls. The admiration which the visitor felt as he passed quietly along near the American Fall was changed into awe when he began to feel the mighty pulse of the great deep just below the tower, then swung round into the white foam directly in front of the Horseshoe, and saw the sky of waters falling toward him. And he seemed to be lifted on wings as he sailed swiftly down on the rushing stream through a baptism of spray. To many persons there was a fascination about it that induced them to make the trip every time they had an opportunity to do so. Owing to some change in her appointments, which confined her to the Canadian shore for the reception of passengers, she became unprofitable. Her owner, having decided to leave the neighborhood, wished to sell her as she lay at her dock. This he could not do, but he received an offer of something more than half of her cost, if he would deliver her at Niagara, opposite the fort. This he decided to do, after consultation with Robinson, who had acted as her captain and pilot on her trips below the Falls. The boat required for her navigation an engineer, who also acted as fireman, and a pilot.

Mr. Robinson agreed to act as pilot for the fearful voyage, and the engineer, Mr. Jones, consented to go with him. A courageous machinist, Mr. McIntyre, volunteered to share the risk with them. They put her in complete trim, removing from deck and hold all superfluous articles. Notice was given of the time for starting, and a large number of people assembled to see the fearful plunge, no one expecting to see the crew again alive after they should leave the dock. This dock, as has been before stated, was just above the Railway Suspension Bridge, at the place where she was built, and where she was laid up in the winter—that, too, being the only place where she could lie without danger of being crushed by the ice. Twenty rods below this eddy the water plunges sharply down into the head of the crooked, tumultuous rapid which we have before noticed as reaching from the bridge to the Whirlpool. At the Whirlpool, the danger of being drawn under was most to be apprehended; in the rapids, of being turned over or knocked to pieces. From the Whirlpool to Lewiston is one wild, turbulent rush and whirl of water, without a square foot of smooth surface in the whole distance.

About three o'clock in the afternoon of June 15, 1861, the engineer took his place in the hold, and, knowing that their flitting would be short at the best, and might be only the preface to swift destruction, set his steam-valve at the proper gauge, and awaited—not without anxiety—the tinkling signal that should start them on their flying voyage. McIntyre joined Robinson at the wheel on the upper deck. Self-possessed, and with the calmness which results from undoubting courage and confidence, yet with the humility which recognizes all possibilities, with downcast eyes and firm hands, Robinson took his place at the wheel and pulled the starting bell. With a shriek from her whistle and a white puff from her escape-pipe, to take leave, as it were, of the multitude gathered on the shores and on the bridge, the boat ran up the eddy a short distance, then swung round to the right, cleared the smooth water, and shot like an arrow into the rapid under the bridge. Robinson intended to take the inside curve of the rapid, but a fierce cross-current carried him to the outer curve, and when a third of the way down it a jet of water struck against her rudder, a column dashed up under her starboard side, heeled her over, carried away her smokestack, started her overhang on that side, threw Robinson flat on his back, and thrust McIntyre against her starboard wheel-house with such force as to break it through. Every eye was fixed, every tongue was silent, and every loooker-on breathed freer as she emerged from the fearful baptism, shook her wounded sides, slid into the Whirlpool, and for a moment rode again on an even keel. Robinson rose at once, seized the helm, set her to the right of the large pot in the pool, then turned her directly through the neck of it. Thence, after receiving another drenching from its combing waves, she dashed on without further accident to the quiet bosom of the river below Lewiston.

Thus was accomplished one of the most remarkable and perilous voyages ever made by men. The boat was seventy-two feet long, with seventeen feet breadth of beam and eight feet depth of hold, and carried an engine of one hundred horse-power. In conversation with Robinson after the voyage, he stated that the greater part of it was like what he had always imagined must be the swift sailing of a large bird in a downward flight; that when the accident occurred the boat seemed to be struck from all directions at once; that she trembled like a fiddle-string, and felt as if she would crumble away and drop into atoms; that both he and McIntyre were holding to the wheel with all their strength, but produced no more effect than they would if they had been two flies;

that he had no fear of striking the rocks, for he knew that the strongest suction must be in the deepest channel, and that the boat must remain in that. Finding that McIntyre was somewhat bewildered by excitement or by his fall, as he rolled up by his side but did not rise, he quietly put his foot on his breast, to keep him from rolling around the deck, and thus finished the voyage.

Poor Jones, imprisoned beneath the hatches before the glowing furnace, went down on his knees, as he related afterward, and although a more earnest prayer was never uttered and few that were shorter, still it seemed to him prodigiously long. To that prayer he thought they owed their salvation.

The effect of this trip upon Robinson was decidedly marked. As he lived only a few years afterward, his death was commonly attributed to it. But this was incorrect, since the disease which terminated his life was contracted at New Orleans at a later day. "He was," said Mrs. Robinson to the writer, "twenty years older when he came home that day than when he went out." He sank into his chair like a person overcome with weariness. He decided to abandon the water, and advised his sons to venture no more about the rapids. Both his manner and appearance were changed. Calm and deliberate before, he became thoughtful and serious afterward. He had been borne, as it were, in the arms of a power so mighty that its impress was stamped on his features and on his mind. Through a slightly opened door he had seen a vision which awed and subdued him. He became reverent in a moment. He grew venerable in an hour.

Yet he had a strange, almost irrepressible, desire to make this voyage immediately after the steamer was put on below the Falls. The wish was only increased when the first *Maid of the Mist* was superseded by the new and stancher one. He insisted that the voyage could be made with safety, and that it might be made a good pecuniary speculation.

He was a character—an original. Born on the banks of the Connecticut, in the town of Springfield, Massachusetts, it was in the beautiful reach of water which skirts that city that he acquired his love of aquatic sports and exercises and his skill in them. He was nearly six feet in stature, with light chesnut hair, blue eyes, and fair complexion. He was a kind-hearted man, of equable temper, few words, cool, deliberate, decided; lithe as a Gaul and gentle as a girl. It goes without saying that he was a man of " undaunted courage." He had that calm, serene, supreme equanimity of temperament which fear could not reach nor disturb. He might have been, under right conditions, a quiet, willing martyr, and at last he bore patiently the wearying hours of slow decay which ended his life. His love of nature and adventure was paramount to his love of money, and although he was never pinched with poverty, he never had abundance.

He loved the water, and was at home in it or on it, as he was a capital swimmer and a skillful oarsman. Especially he delighted in the rapids of the Niagara. Kind and compassionate as he was by nature, he was almost glad when he heard that a fellow-creature was, in some way, entangled in the rapids, since it would give him an excuse, an opportunity, to work in them and to help him. As he was not a boaster, he made no superfluous exhibitions of his skill or courage, albeit he might occasionally indulge—and be indulged—in some mirthful manifestation of his good-nature; as when, on reaching Chapin's refuge for his rescue, he waved from one of its tallest cedars a green branch to the anxious spectators, as if to assure and encourage them; and when he returned with his skiff half filled with cedar-sprigs, which he distributed to the multitude, they raised his pet craft to their shoulders, with both Chapin and himself in it, and bore them in triumph through the village, while money tokens were thrown into the boat to replace the green ones.

He never foolishly challenged the admiration of his fellow-men. But when the emergency arose for the proper exercise of his powers, when news came that some one was in trouble in the river, then he went to work with a calm and cheerful will which gave assurance of the best results. Beneath his quiet deliberation of manner there was concealed a wonderful vigor both of resolution and nerve, as was amply shown by the dangers which he faced, and by the bend in his withy oar as he forced it through the water, and the feathery spray which flashed from its blade when he lifted it to the surface.

In all fishing and sailing parties his presence was indispensable for those who knew him. The most timid child or woman no longer hesitated if Robinson was to go with the party. His quick eye saw everything, and his willing hand did all that it was necessary to do, to secure the comfort and safety of the company.

It is doubtful whether more than a very few of his neighbors know where he lies, in an unmarked grave in Oakwood Cemetery, near the rapids. Robinson went forth on a turbulent, unreturning flood, where the slightest hesitancy in thought or act would have proved instantly fatal. Benevolent associations in different cities and countries bestow honor and rewards on those who, by unselfish effort and a noble courage, save the life of a fellow-being. This Robinson did repeatedly, yet no monument commemorates his worthy deeds.

CHAPTER XII.

A fisherman and a bear in a canoe — Frightful experience with floating ice — Early farming on the Niagara — Fruit growing — The original forest — Testimony of the trees — The first hotel — General Whitney — Cataract House — Distinguished visitors — Carriage road down the Canadian bank — Ontario House — Clifton House — The Museum — Table and Termination Rocks — Burning Spring — Lundy's Lane — Battle Anecdotes.

SOON after the War of 1812, a fisherman — whose name we will call Fisher — on a certain day went out upon the river, about three miles above the Fall; and while anchored and fishing from his canoe, he saw a bear in the water making, very leisurely, for Navy Island. Not understanding thoroughly the nature and habits of the animal, thinking he would be a capital prize, and having a spear in the canoe, he hoisted anchor and started in pursuit. As the canoe drew near, the bear turned to pay his respects to its occupant. Fisher, with his spear, made a desperate thrust at him. Quicker and more deftly than the most expert fencer could have done it, the quadruped parried the blow, and, disarming his assailant, knocked the spear more than ten feet from the canoe. Fisher then seized a paddle and belabored the bear over his head and on his paws, as he placed the latter on the side of the canoe and drew himself in. The

Fisher and the Bear.

Opposite page 97.

now frightened fisherman, not knowing how to swim, was in a most uncomfortable predicament. He felt greatly relieved, therefore, when the animal deliberately sat himself down, facing him, in the bow of the canoe. Resolving in his own mind that he would generously resign the whole canoe to the creature as soon as he should reach the land, he raised his paddle and began to pull vigorously shoreward, especially as the rapids lay just below him, and the Falls were roaring most ominously.

Much to his surprise, as soon as he began to paddle Bruin began to growl, and, as he repeated his stroke, the occupant of the bow raised his note of disapproval an octave higher, and at the same time made a motion as if he would attack him. Fisher had no desire to cultivate a closer intimacy, and so stopped paddling.

Bruin serenely contemplated the landscape in the direction of the island. Fisher was also intensely interested in the same scene, and still more intensely impressed with their gradual approach to the rapids. He tried the paddle again. But the tyrant of the quarter-deck again emphatically objected, and as *he* was master of the situation, and fully resolved not to resign the command of the craft until the termination of the voyage, there was no alternative but submission. Still, the rapids were frightfully near and something must be done. He gave a tremendous shout. But Bruin was not in a musical mood, and vetoed that with as much emphasis as he had done the paddling. Then he turned his eyes on Fisher quite interestedly, as if he were calculating the best method of dissecting him. The situation was fast becoming something more than painful. Man and bear in opposite ends of the canoe floating—not exactly double—but together to inevitable destruction. But every suspense has an end. The single shout, or something else, had called the attention of the neighbors to the canoe. They came to the rescue, and an old settler, with a musket which he had used in the War of 1812, fired a charge of buck-shot into Bruin which induced him to take to the water, after which he was soon taken, captive and dead, to the shore. He weighed over three hundred pounds.

A son of the settler who shot the bear had a frightful experience in the river many years afterward. He was engaged in Canada in the business of buying saw-logs for the American market. Coming from the woods down to Chippewa one cold day in December, at a time when considerable quantities of strong, thin cakes of ice were floating in the river, he took a flat-bottom skiff to row across to his home. This he did without apprehension, as he had been born and brought up on the banks of the Niagara, understood it well, and was also a strong, resolute man.

As he drew near the foot of Navy Island, intending to take the chute between it and Buckhorn Island, two large cakes between which he was sailing suddenly closed together and cut the bottom of his skiff square off. Just above the cake on which his bottomless skiff was then floating there was a second large cake, at a little distance from it, and beyond this a strip of water which washed the shore of Navy Island. In less time than it has taken to write this, he sprang upon the first piece of ice, ran across it with desperate speed, cleared the first space of water at a single leap, ran across the next cake of ice, jumped with all his might, and landed in the icy water within a rod of the shore, to which he swam. He was soon after warming and drying himself before the rousing fire of the only occupant of the island.

His father had a fine farm on the bank of the river, which he cultivated with much care. But before the drainage of the country was completed the land was decidedly wet. A friend from the East who made him a call found him plowing. The water stood in the bottom of the furrows. But agriculture has been progressive since those days. It is now almost a fine art instead of a mere pursuit. And nowhere north of the equator is there a climate and soil so genial and favorable for the growth of certain kinds of fruit, especially the apple and the peach, as are those of Niagara County. Many persons claim that they can tell from the peculiar consistency of the pulp, and by its flavor and *bouquet*, on which side of the Genesee River an apple is grown.

It is said that the winter apples of Niagara are as well known and as greatly prized above all others of their kind on the docks of Liverpool, as is Sea Island cotton above all other grades of that plant. The delicious little russet known as the *Pomme Gris*, with its fine aromatic flavor when ripe, grows nowhere else to such perfection as along the Niagara River. In 1825, at the grand celebration held to commemorate the completion of the

Erie Canal, the late Judge Porter made the first shipment east of apples raised in Niagara County. It consisted of two barrels, one of which was sent to the corporation of the city of Troy, and the other to that of New York. They were duly received and honored. From this small beginning the fruit trade has grown to the yearly value of more than a million of dollars for Niagara County alone.

With reference to the forest which once covered this country, an erroneous impression prevails as to its age. Poets and romancers have been in the habit of speaking of these "primeval forests" as though they might have been bushes when Nahor and Abraham were infants. But this is a great error. Since the discovery of the country only one tree has been found that was eight hundred years old. This is mentioned by Sir Charles Lyell as having grown out of one of the ancient mounds near Marietta, Ohio. But the great majority of them were not over three hundred years old. The testimony of the trees concerning the past is not quite so abundant as that of the rocks, but that of one tree grown in central New York is of a remarkable character. It was a white oak, which grew in the rich valley of the Clyde River, about one mile west of Lyons' Court House, and was cut down in the year 1837. The body made a stick of timber eighty feet long, which before sawing was about five feet in diameter. It was cut into short logs and sawed up. From the center of the butt-log was sawed a piece about eight by twelve inches. At the butt end of this piece the saw laid bare, without marring them, some old scars made by an ax or some other sharp instrument. These scars were perfectly distinct and their character equally unmistakable. They were made, apparently, when the young tree was about six inches in diameter. Outside of these scars there were counted four hundred and sixty distinct rings, each ring marking with unerring certainty one year's growth of the tree. It follows that this chopping was done in 1374, or one hundred and eighteen years before the first voyage of Columbus across the Atlantic.

It has been questioned whether the rings shown in a cross-section of a tree can be relied upon to determine truly the number of years it has been growing. A singular confirmation of the correctness of this method of counting was furnished some years since.

In the latter part of the last century the late Judge Porter surveyed a large tract of land lying east of the Genesee River, known as "The Gore." Some thirty-five years afterward it became necessary to resurvey one of its lines, and recourse was had to the original surveys. Most of the forest through which the first line had been run was cleared off, and such trees as had been "blazed" as line-trees had overgrown the scars. One tree was found which was declared to be an original line-tree. On cutting into it carefully the old "blaze" was brought to light, and on counting the rings outside of it, they were found to correspond with the number of years which had elapsed since the first survey.

One of the three small buildings at Niagara which escaped the flames of 1814 was a log-cabin, about thirty by forty feet in its dimensions, that stood in the center of the front of the International block. In the latter part of 1815 the inhabitants returned, and the late General P. Whitney put a board addition to the log-house, and opened the first hotel. From that has grown up the present International. The immediate predecessor of the International was the Eagle Tavern, which was, for some years, in charge of a genial and popular landlord, the late Mr. Hollis White. It was formed by the addition to the old frame structure of a three-story brick building, of moderate dimensions. Across the front of this addition was a long, wide, old-fashioned stoop. This was well supplied with comfortable arm-chairs, which furnished easy rests for guests or neighbors, and were well patronized by both, and especially during the summer season by the genial humorists of the place. On the opposite side of the street was a small house, a story and a half high, belonging to Judge Porter, and to which he built an addition. Then, as now, there were occasionally more visitors than the hotel could accommodate, and the neighbors assisted in entertaining them. Judge Porter did this frequently, and among his guests were President Monroe, Marshal Grouchy, General La Fayette, General Brown, General Scott, Judge Spencer, and other distinguished strangers.

The first building erected on the ground where the Cataract House now stands was of a later date—1824—a frame house about fifty feet square. It was purchased by General Whitney in 1826, and formed the nucleus of the great pile which constitutes the present Cataract House.

In 1829, the carriage road down the bank to the ferry on the Canadian side was made. For several years previous the principal hotel at the Falls was also on that side. It was called the Pavilion, and stood on the high bank just above the Horseshoe Fall. It commanded a grand view of the river above, and almost a bird's-eye view of the Falls and the head of the chasm below. The principal stage-route from Buffalo was likewise on that side, and the register of the Pavilion contained the names of most of the noted visitors of the period. But the erection of the Cataract House and the establishing of stage-routes on the American side drew away much of its patronage, and finally, on the completion of the first half of the Clifton House, in 1833, it was quite abandoned. A few years later the Ontario House was built, about half-way between the Clifton and the Horseshoe Fall, toward which it fronted. There was not sufficient business to support it, and after standing unoccupied for several years, it took fire and was burned to the ground.

The Clifton was greatly enlarged and improved by Mr. S. Zimmerman in 1865. The Amusement Hall and several cottages were built and gas-works erected. The grounds were handsomely graded and adorned.

Near the site of Table Rock is the Museum, its valuable collection being the result of several years' labor by its proprietor, Mr. Thomas Barnett. It contains several thousand specimens from the animal and mineral kingdoms, and the galleries are arranged to represent a forest scene.

Just above the Museum the visitor steps upon what

remains of the famous Table Rock. It was once a bare rock pavement, about fifteen rods long and about five rods wide, about fifty feet of its width projecting beyond its base at the bottom of the limestone stratum nearly one hundred feet below. Remembering this fact, we can more readily credit the probable truth of the statement made by Father Hennepin—which we have before noticed—that the projection on the American side in 1682, when he returned from his first tour to the West, was so great that four coaches could drive abreast under it. On top of the *débris* below the bank lies the path by which Termination Rock, under the western end of the Horseshoe, is reached. It is a path which few neglect to follow.

The Table itself has always been, and must continue to be, a favorite resort for visitors. The combined view of the Falls and the chasm below, as well as the rapids above, is finer, more extensive, here than from any other point. Moreover, the nearness to the great cataract is more sensibly felt, the communion with it is deeper and more intimate than it can be anywhere else. The view from this point can be most pleasantly and satisfactorily taken in the afternoon, when the spectator has the sun behind him, and can look at his leisure and with unvexed eyes at the brilliant scene before him. However long he may tarry he will find new pleasure in each return to it.

Two miles above, following round the bend of the Oxbow toward Chippewa, and down near the water's edge, is the Burning Spring. The water is impregnated with sulphureted hydrogen gas, and is in a constant state of mild ebullition. The gas is perpetually rising to the surface of the water, and when a lighted match is applied it burns with an intermittent flame. If, however, a tub with an iron tube in the center of its bottom is placed over the spring, a constant stream of gas passes through it. On being lighted it burns constantly, with a pale blue, wavering flame, which possesses but little illuminating or heating power. The drive is a pleasant one, affording a fine view of the Oxbow Rapids and islands and the noble river above.

A mile and a quarter west of Table Rock is the Lundy's Lane battle-ground. On the crown of the hill, where the severest struggle occurred, are two rival pagodas challenging the tourist's attention. From the top of each he has a rare outlook over a broad level plain, relieved on its northern horizon by the top of Brock's Monument, and to the south-east by the city of Buffalo and Lake Erie.

The obliging custodian of either tower will enlighten his hearers with dextrous volubility, and, according as he is certain of the nationality of his listeners, will the Stars and Stripes wave in triumph, or the Cross of Saint George float in glory, over the bloody and hard-fought field. If he cannot feel sure of his listeners' habitat, like Justice, he will hold an even balance and be blind withal.

It was the writer's privilege to go over the field on a pleasant June day with Generals Scott and Porter, and to learn from them its stirring incidents. General Scott pointed out the location of the famous battery on the

British left which made such havoc with his brave brigade, and in taking which the gallant Miller converted his modest "I'll try, sir," into a triumphant "It is done." The General also found the tree under which, faint from his bleeding wound, he sat down to rest, placing its protecting boll between his back and the British bullets, as he leaned against it. Plucking a small wild flower growing near it, he presented it to one of the ladies of the party, telling her that "it grew in soil once nourished by his blood."

General Porter showed us where, with his volunteers and Indians, he broke through the woods on the British right, just as Miller had captured the troublesome battery, thus aiding to win the most obstinate and bloody fight of the war. Its hard-won trophies, however, were too easily lost, as, by some misunderstanding or neglect of orders, the proper guard around the field was not maintained, and, in the darkness proverbially intense just before day, the British returned to the field and quietly removed most of the guns. So our English friends claim it was a drawn battle.

Nearly half a century later a dinner was given at Queenston by our Canadian friends, to signalize the completion of the Lewiston Suspension Bridge. On this occasion a British-Canadian officer, the late Major Woodruff, of St. David's, who served with his regiment during the war, was called upon by the chairman, the late Sir Allan McNabb, to follow, in response to a toast, the late Colonel Porter, only son of General Porter. In a mirthful reference to the stirring events of the war he alluded to the British retreat after the battle of Chippewa, and condensing the opposing forces into two personal pronouns, one representing General Porter and the other himself, he turned to Colonel Porter and said: "Yes, sir, I remember well the *moving* events of that day, and how sharp he was after me. But, sir, he was balked in his purpose, for although he won the *victory* I won the *race*, and so we were even."

CHAPTER XIII.

Incidents — Fall of Table Rock — Remarkable phenomenon in the river — Driving and lumbering on the Rapids — Points of the compass at the Falls — A first view of the Falls commonly disappointing — Lunar bow — Golden spray — Gull Island and the gulls — The highest water ever known at the Falls — The Hermit of the Falls.

O F incidents, curious, comic, and tragic, connected with the locality the catalogue is long, but we must make our recital of them brief.

We have before referred to Professor Kalm's notice of the fall of a portion of Table Rock previous to 1750. Authentic accounts of like events are the following: In 1818 a mass one hundred and sixty feet long by thirty wide; in 1828 and '29 two smaller masses; also in 1828 there went down in the center of the Horseshoe a huge mass, of which the top area was estimated at half an acre. If this estimate was correct, it would show an abrasion equivalent to nearly one foot from the whole surface of the Canadian Fall. In April, 1843, a mass of rock and earth about thirty-five feet long and six feet wide fell from the middle of Goat Island. In 1847, just north of the Biddle Stairs, there was a slide of bowlders, earth, and gravel, with a small portion of the bed-rock, the whole mass being about forty feet long and ten feet wide. About

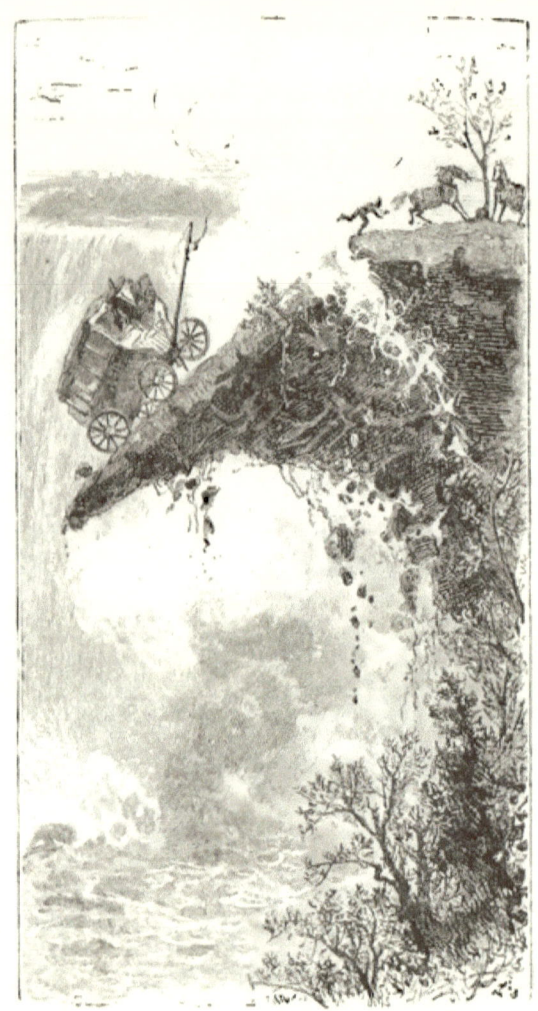

Opposite page 119. Fall of Table Rock.

every third return of spring has increased the abrasion at these two points. At the first-named point more than twenty feet in width has disappeared, with the whole of the road crossing the island. From the latter point, near the Biddle Stairs, which was a favorite one for viewing the Horseshoe Fall, the seats provided for visitors and the trees which shaded them have fallen.

On the 25th of June, 1850, occurred the great downfall which reduced Table Rock to a narrow bench along the bank. The portion which fell was one immense solid rock two hundred feet long, sixty feet wide, and one hundred feet deep where it separated from the bank. The noise of the crash was heard like muffled thunder for miles around. Fortunately it fell at noonday, when but few people were out, and no lives were lost. The driver of an omnibus, who had taken off his horses for their midday feed, and was washing his vehicle, felt the preliminary cracking and escaped, the vehicle itself being plunged into the gulf below.

In 1850, a canal-boat that became detached from a raft, went down the Canadian Rapids, turned broadside across the river before reaching the Falls, struck amidships against a rock projecting up from the bottom and lodged. It remained there more than a year, and when it went down took with it a piece of the rock apparently about ten feet wide and forty feet long. At the foot of Goat Island some smaller masses have fallen, and three extensive earth-slides have occurred.

In the spring of 1852 a triangular mass, the vertex of which was just beyond or south of the Terrapin Tower, while its altitude of more than forty feet lay along the shore of the south corner of Goat Island, fell in the night with the usual grinding crash. And with it fell some isolated rocks which lay on the brink of the precipice in front of the tower, and from which the tower derived its name. Before the tower was built, some person looking at the rocks from the shore suggested that they looked like huge terrapins sunning themselves on the edge of the Fall. A few days after the fall of the triangular mass, a huge column of rock a hundred feet high, about fourteen feet by twelve, and flat on the top, became separated from the bank and settled down perpendicularly until its top was about ten feet below the surface rock. It stood thus about four years, when it began gradually to settle, as the shale and stone were disintegrated beneath it, and finally it tumbled over upon the rocks below, furnishing an illustration of the manner in which we suppose the rocks which once accumulated below the Whirlpool must have been broken down. In the spring of 1871 a portion of the west side of the sharp angle of the Horseshoe, apparently about ten by thirty feet, went down, producing a decided change in the curve.

On the 7th day of February, 1877, about eleven o'clock of a cold, cloudy day, there occurred the most extensive abrasion of the Horseshoe Fall ever noted. It extended from near the water's edge at Table Rock, more than half the distance round the curve, some fifteen hundred feet, and at the most salient angle the mass that fell was from fifty to one hundred feet wide. By this downfall the contour of the Horseshoe was decidedly changed, the reëntering angle being made acute and very ragged. Less than three months afterward the abrasion was continued some two hundred feet toward Goat Island.

The trembling earth and muffled thunder gave evidence of the immensity of the mass of fallen rock, but no one saw it go down. For several months after the fall, until the mass of rock got thoroughly settled in the bed of the Falls, the exhibition of water-rockets, sent up a hundred feet above the top of the precipice, was unique and beautiful. The greatest angle of retrocession, which had previously been wearing toward Goat Island, is again turning toward the center of the stream.

On the 29th of March, 1848, the river presented a remarkable phenomenon. There is no record of a similar one, nor has it been observed since. The winter had been intensely cold, and the ice formed on Lake Erie was very thick. This was loosened around the shores by the warm days of the early spring. During the day, a stiff easterly wind moved the whole field up the lake. About sundown, the wind chopped suddenly round and blew a gale from the west. This brought the vast tract of ice down again with such tremendous force that it filled in the neck of the lake and the outlet, so that the outflow of the water was very greatly impeded. Of course, it only needed a short space of time for the Falls to drain off the water below Black Rock.

The consequence was that, when we arose in the morning at Niagara, we found our river was nearly half gone. The American channel had dwindled to a respectable creek. The British channel looked as though it had been smitten with a quick consumption, and was fast passing away. Far up from the head of Goat Island and out into the Canadian rapids the water was gone, as it was also from the lower end of Goat Island, out beyond the tower. The rocks were bare, black, and forbidding. The roar of Niagara had subsided almost to a moan. The scene was desolate, and but for its novelty and the certainty that it would change before many hours, would have been gloomy and saddening. Every person who has visited Niagara will remember a beautiful jet of water which shoots up into the air about forty rods south of the outer Sister in the great rapids, called, with a singular contradiction of terms, the "Leaping Rock." The writer drove a horse and buggy from near the head of Goat Island out to a point above and near to that jet. With a log-cart and four horses, he drew from the outside of the outer island a stick of pine timber hewed twelve inches square and forty feet long. From the top of the middle island was drawn a still larger stick, hewed on one side and sixty feet long.

There are few places on the globe where a person would be less likely to go lumbering than in the rapids of Niagara, just above the brink of the Horseshoe Fall. All the people of the neighborhood were abroad, exploring recesses and cavities that had never before been exposed to mortal eyes. The writer went some distance up the shore of the river. Large fields of the muddy bottom were laid bare. The shell-fish, the uni-valves, and the bi-valves were in despair. Their housekeeping and domestic arrangements were most unceremoniously exposed. The clams, with their backs up and their open mouths down in the mud, were making their sinuous courses toward the shrunken stream. The small-fry of fishes were wriggling in wonder to find themselves impounded in small pools.

This singular syncope of the waters lasted all the day, and night closed over the strange scene. But in the morning our river was restored in all its strength and beauty and majesty, and we were glad to welcome its swelling tide once more.

It is a curious fact that nine out of every ten persons who visit the Falls for the first time, are on their arrival completely bewildered as to the points of the compass; and this without reference to the direction from which they may approach them. All understand the general geographical fact that Canada lies north of the United States. Hence they naturally suppose, when they arrive at the frontier, that they must see Canada to the north of them. But when they reach Niagara Falls they look across the river into Canada, in one direction directly south, and in another directly west. Only a reference to the map will rectify the erroneous impression. It is corrected at once by remembering that the Niagara River empties into the south side of Lake Ontario.

One other fact may be regarded as well-established, namely, that most visitors are disappointed when they first look upon the Falls. They are not immediately and forcibly impressed by the scene, as they had expected to be. The reasons for this are easily explained. The chief one is that the visitor first sees the Falls from a point above them. Before seeing them, he reads of their great height; he expects to look up at them and behold the great mass of water falling, as it were, from the sky. He reads of the trembling earth; of the cloud of spray, that may be seen a hundred miles away; of the thunder of the torrent, and of the rainbows. He does not consider that these are occasional facts. He may not know he is near the Falls until he gets just over them. At certain times he feels no trembling of the earth; he hears no stunning roar; he may see the spray scattered in all directions by the wind, and of course he will see no bow. Naturally, he is disappointed. But it is not long before the grand reality begins to break upon him, and every succeeding day and hour of observation impresses him more and more deeply with the vastness, the power, the sublimity of the scene, and the wonderful and varied beauty of its surroundings. Those who spend one or more seasons at Niagara know how very little can be seen or comprehended by those who "stop over one train."

They are fortunate who can see the Falls first from the ferry-boat on the river below, and about one-third of the way across from the American shore. The writer has frequently tried the experiment with friends who were willing to trust themselves, with closed eyes, to his guidance, and wait until he had given them the signal to look upward.

Those who may be at Niagara a few nights before and

Rock of Ages and Whirlwind Bridge.
Opposite page 114.

after a full moon should not fail to go to Goat Island to see the lunar bow. It is the most unreal of all real things — a thing of weird and shadowy beauty.

Another striking scene peculiar to the locality is witnessed in the autumn, when the sun in making its annual southing reaches a point which, at the sunset hour, is directly west from the Falls. Then those who are east of them see the spray illuminated by the slant rays of the sinking sun. In the calm of the hour and the peculiar atmosphere of the season, the majestic cloud looks like the spray of molten gold.

In 1840 there was a small patch of stones, gravel, sand, and earth, called Gull Island, lying near the center of the Canadian rapid and about one hundred rods above the Horseshoe Fall. It was apparently twenty rods long by two rods wide, and was covered with a growth of willow bushes. It was so named because it was a favorite resort of that singular combination of the most delicate bones and lightest feathers called a gull.

The birds seem large and awkward on the wing, but as they sit upon the water nothing can appear more graceful. They are far-sighted and keen-scented. Their eyes are marvels of beauty. They are eccentric in their habits, the very Arabs of their race — here to-day and gone to-morrow. They are gregarious and often assemble in large numbers. At times in a series of wild, rapid, devious gyrations, and uttering a low, mournful murmur, they seem to be engaged, as it were, in some solemn festival commemorative of their departed kindred. One moment the air will be filled with them and their sad refrain; the next moment the cry will have ceased and not a gull will be seen. They come as they go, summer and winter alike. In thirty years the writer has never been able to discover when nor whence they came. In winter they generally appear in the milder days, and their disappearance is followed by cooler weather.

In the spring of 1847 a long and fierce gale from the west, which drove the water down Lake Erie, caused the highest rise ever known in the river. It rose six feet on the rapids, and for the first time reached the floor-planking of the old bridge. The greater part of Gull Island was washed down in this flood, and ten years later it had wholly disappeared.

The vague tradition—the origin of which cannot be traced—that there is a flux and reflux of the waters in the Great Lakes, which embraces a period of about seven years, is not confirmed by our observation, if it be intended to affirm that the ebb and flow are both completed in seven years. Our observation shows that there is a flow of about seven years, and a reflux, which is accomplished in the same period. The water in the Niagara was very low in 1843-4, again in 1857-8, and again in 1871-2. This last is the lowest long continued shrinkage ever known. It is, however, altogether probable that the general level of the lakes will fall hereafter, owing to the destruction of the forests and the cultivation of the land along their shores. In this case the waters of the Niagara and Detroit rivers may, in the far future, meet in the bed of Lake Erie, and their margins be covered with orchards and vineyards more extensive and productive than those along the Rhine.

The Hermit of the Falls, so called, Mr. Francis Abbott, came to the village in June, 1829. He was a rather good-looking, respectable young man, of moderate attainments, who was subject, apparently, to a mild form of intermittent derangement. Though his manner was eccentric, his conduct was harmless, and it is probable that his parents, who, it was afterward ascertained, were respectable members of the Society of Friends in England, encouraged his desire to travel, and furnished him the means to do so. He seems to have had some taste for music, and to have been a tolerable performer on the flute. He wandered much about the island, both night and day, and often bathed below the little fall on the south side of Goat Island, near its head. He lived alone in an unoccupied log-hut, directly across the island from this fall, until about the first of April, 1831, when he removed to a little cabin of his own building, on Point View. In June of that year, just two years after his arrival, he was drowned while bathing below the ferry. Ten days after, his body was found at Fort Niagara, brought back, and buried in the God's-acre at the Falls.

CHAPTER XIV.

Avery's descent of the Falls — The fatal practical joke — Death of Miss Rugg — Swans — Eagles — Crows — Ducks over the Falls — Why dogs have survived the descent.

ON the morning of the 19th of July, 1853, a man was discovered in the middle of the American rapid, about thirty rods below the bridge. He was clinging to a log, which the previous spring had lodged against a rock. He proved to be a Mr. Avery, who had undertaken to cross the river above the night before, but, getting bewildered in the current, was drawn into the rapids. His boat struck the log, and was overturned, yet, by some extraordinary good fortune, he was able to hold to the timber. A large crowd soon gathered on the shore and bridge. A sign, painted in large letters, "We will save you," was fastened to a building, that the reading of it might cheer and encourage him. Boats and ropes were provided, with willing hands to use them. The first boat lowered into the rapids filled and sank just before reaching Avery. The next, a life-boat, which had been procured from Buffalo, was let down, reached the log, was dashed off by the reacting waters, upset, and sank beside him. Another light, clinker-built boat was launched, and reached him just right. But, in some unaccountable manner, the rope got caught between the rock and the log. It was impossible to loosen it. Poor Avery tugged and worked at it with almost superhuman energy for hours. The citizens above pulled at the rope until it broke.

By this time a raft had been constructed, with a strong cask fastened to each corner, and ropes attached so that Avery could tie himself to it. It was lowered, and reached him safely. He got on it and seized the ropes. Every heart grew lighter as the rescuers moved across the lower part of Bath Island, drawing in the rope, while the raft swung easily toward Goat Island. But when it reached the head of Chapin's Island, all hopes were dashed again. The rope attached to the raft got caught in the rocks as it was passing below a ledge in a swift chute of water. All efforts to loosen it were ineffectual. Another boat was launched and let downstream. It reached the raft all right, and Avery, in his eagerness to seize it, dropped the ropes he had been holding, stepped to the edge of the raft, with his hands extended to catch the boat, when the raft, under his weight, settled in the water, and, just missing his hold, he was swept into the rapids, went down the north side of Chapin's Island, and, almost in reach of it, in water so shallow that he regained his feet for an instant, threw up his hands in despair, fell backward, and went over the Fall. The tragedy lasted eighteen hours.

The names connected with the next incident are suppressed, out of regard for the feelings of surviving friends. It is given as a warning to future visitors to Niagara not to attempt any mirthful experiments around the Falls.

A party of ladies, gentlemen, and children were on Luna Island, near a small beech tree, since destroyed, called "the Parasol." A young girl of ten was standing near her mother, just on the brink of the water, when a young man of twenty-two stepped up beside her and seized her playfully by the arms, saying, "Now, Nannie, I am going to throw you in," and swung her out over the water. Taken by surprise and frightened, she struggled, twisted herself out of his grasp, and fell into the rapid within twenty feet of the brink of the precipice. Instantly the young man plunged in after her, seized hold of her dress, and swung her around toward her half-distracted mother, who almost reached her as she slipped by and went over the Fall, immediately followed by the young man. The young girl was found some days afterward, lying on her back, on a large rock, holding her open parasol above her head, as though she had lain down to rest. A few weeks afterward the father of the young man was coming up the river, on the *Maid of the Mist*, from the lower landing. A body was discovered floating in the water, and, by the aid of a small boat, was brought on board the steamer. It was that of his son.

On the 23d of August, 1844, Miss Martha K. Rugg was walking to Table Rock with a friend. Seeing a bunch of cedar-berries on a low tree, which grew out from the edge of the bank, she left her companion, reached out to pick it, lost her footing, and fell one hundred and fifteen feet upon the rocks below. She survived about three hours. Pilgrims to Table Rock used to inquire for the spot where this accident happened. The following spring, an enterprising Irishman brought out a table of suitable dimensions, set it down on the bank of the river, and covered it with different articles, which he offered for sale. In order to enlighten strangers about the spot, he provided a remarkable sign, which he set up near one end of the table. This sign was a monumental obelisk, about five feet high, made of pine boards, and painted white. On the base he painted, in black letters, the following inscription :

> " Ladies fair, most beauteous of the race,
> Beware and shun a dangerous place.
> Miss Martha Rugg here lost a life,
> Who might now have been a happy wife."

An envious competitor, one of his own countrymen, brought his own table of wares, and placed it just above the original mourner. Thereupon, the latter, determining that his rival should not have the benefit of his sign, removed it below his own table, having first removed the table itself as far down as circumstances would permit. Then he added his master-stroke of policy. Up to that time the monument had been stationary. Thenceforward, every day on quitting business he put it on a wheelbarrow and took it home, bringing it out again on resuming operations in the morning.

Previous to the War of 1812, the Niagara River abounded in swans, wild geese, and ducks. Since that war none of the swans have been seen here, except two pair which came at different times. One of each pair went over the Falls, and was taken out alive but stunned. Their mates, faithful unto death, were shot while watching and waiting for their return.

Eagles have always been seen in the vicinity, and a few have been captured. A single pair for many years had their aerie in the top of a huge dead sycamore tree, near the head of Burnt Ship Bay. It was interesting to watch the flight of the male bird when he left his brooding mate to go on a foraging expedition. Leaving the topmost limb that served as his home observatory, he would sweep round in a circle, forming the base of a regular spiral curve, in which he rose to any desired height. Then, having apparently determined by scent or sight, or by both, the direction he would take, he sailed grandly off. How grandly, too, on his return, he floated to his lofty perch with a single fold of his great wings, and sat for a few moments, motionless as a statue, before greeting his mate. When the young eaglets had but recently chipped their shells, passing sportsmen were content to view the majestic pair at a respectful distance. A pair of eagles, each carrying ten talons, a hooked beak, a strong pair of wings, and an unerring eye, all backed and propelled by an indomitable will and courage, are not to be recklessly trifled with.

Early in July, 1877, two farmers riding in a buggy from Bergholtz, in the easterly part of the town of Niagara, toward the town of Wilson on Lake Ontario, saw a large gray eagle sitting on a fence by the roadside, and watching with much interest some object in a field beyond. Leaving their buggy, they ascertained that the object of its solicitude was an eaglet sitting on the ground, unable to fly, his wings and feathers having been drenched by a heavy shower. One of the men who first reached the young bird found it rather bellicose, and while attempting to secure it was surprised by a vigorous thump on the head from the old bird, accompanied with a sensation of sharp claws in his hair which nearly prostrated him. His assailant then rose quickly some forty feet in the air, and, turning again, descended upon the man with such force as to compel him to relinquish his game. His friend joined him, and for nearly half an hour the two were engaged in a fierce fight with the resolute bird, which they estimated would measure eight feet across the extended wings. The eagle would soar quickly upward as at first until it reached the desired range, when it would turn upon them with great fierceness, thumping with its wings and striking with its talons at their very faces. Finally, securing a number of good-sized cobble-stones, they advanced again upon the eaglet, and were at once attacked by the parent. But they used their stone artillery with vigor, and succeeded in getting the eaglet to their buggy, leaving its gallant defender still unconquered and soaring in the air with a slightly injured wing.

Before the War of the Rebellion, Niagara was a favorite resort of that winged scavenger, the crow, and, at times, they were very numerous. But after the first year of the war they entirely disappeared. Snuffing the battle from afar, they turned instinctively to the South, and did not re-appear among us until several years after the war had ended.

Large numbers of ducks formerly went over the Falls, but not for the reason generally assigned, namely, that they cannot rise out of the rapids. It is true that they cannot rise from the water while heading up-stream.

When they wish to do so, they turn down the current, and sail out without difficulty. No sound and living duck ever went over the precipice by daylight. Dark and especially foggy nights are most fatal to them. In the month of September, 1841, four hundred ducks were picked up below the Falls, that had gone over in the fog of the previous night. In two instances, dogs have been sent over the Falls and have survived the plunge. In 1858 a bull-terrier was thrown into the rapids, also near the middle of the bridge. In less than an hour he came up the ferry-stairs, very wet and not at all gay.

The reason why the dogs were not killed may be thus explained. From the top of the Rapids Tower, before its destruction, the spectator could get a perfect view of the Canadian Fall. On a bright day, by looking steadily at the bottom of the Horseshoe, where water falls into water, he could see, as the spray was occasionally removed, a beautiful exhibition of water-cones, apparently ten or twelve feet high. These are formed by the rapid accumulation and condensation of the falling water. It pours down so rapidly and in such quantities that the water below, so to speak, cannot run off fast enough, and it piles up as though it were in a state of violent ebullition. These cones are constantly forming and breaking. If any strong animal should fall upon one of these cones, as upon a soft cushion, it might slide safely into the current below. The dogs were, doubtless, fortunate enough to fall in this way, aided also by the repulsion of the water from the rocks in the swift channel through which they passed.

The Three Sisters, or Moss Islands.

opposite page 125.

CHAPTER XV.

Wedding tourists at the Falls — Bridges to the Moss Islands — Railway at the ferry — List of persons who have been carried over the Falls — Other accidents.

FOR many years Niagara has been a favorite resort for bridal tourists, who in a crowd of strangers can be so excessively proper that every one else can see how charmingly improper they are.

The three fine, graceful bridges which unite Goat Island with the three smaller islands — the Moss Islands, or the Three Sisters — lying south of it were built in 1858. They opened up a new and attractive feature of the locality, with which all visitors are charmed. Those who have been on them will remember what a broken, wild, tangled mass of rocks, wood, and vines they are. Nothing on Onalaska's wildest shore could be more thoroughly primitive.

A rude path with steps cut in the slope of the bank was for several years the only way of getting down to the water's edge at the ferry. In 1825 several flights of stairs were erected, with good paths between, which made the task quite safe and easy. The double railway-track at the ferry was completed in 1845. When the necessary excavations were nearly finished, and people were told the object of it, the scheme met no approval from those conservative persons who have no faith in new things. The idea of a railway "to go by water" was not considered a brilliant one. Indeed, the greater number shrugged their shoulders at the thought of riding down *that* hill. But as soon as the lumber cars were started for the convenience of the workmen, and people saw how expeditious and easy was the trip, it was difficult to keep them off the cars. Hundreds of thousands of passengers have ridden in them without accident or injury. The motive power is a reaction water-wheel set in a deep pit, and as all the machinery is concealed, it has quite the appearance of a self-working apparatus. There is alongside of the railroad a straight stair-way of two hundred and ninety steps, for those who prefer to use it.

The number of victims whom carelessness or folly has sent over the Falls is large, and, it may be believed, is quite independent of the Indian tradition that the great cataract demands a yearly sacrifice of two human victims.

OVER THE FALLS.

In 1810 the boat *Independence*, laden with salt, filled and sunk while crossing to Chippewa. The captain and two of the crew went over the Falls. One of the crew clung to a large oar, and was saved by a small boat from Chippewa.

1821 Two men in a scow were driven down the current by the wind, and went over the Falls.

1825 Two men in a boat from Grand Island went over.
— Three men went over in three different canoes.
1841 Two men, engaged in smuggling, were upset in the current; one went over. One was found dead on Grass Island.
— Two men who were carrying sand in a scow were drawn into the current and went over.
1847 A lad of fourteen undertook to row across on a Sunday morning, and went over.
1848 In August, a man in a boat passed under the Goat Island Bridge, within ten feet of the shore; he asked of persons on the bridge, "Can I be saved?" Soon after the boat upset, and he went over, feet foremost, struck on the rocks below, and was never seen afterward.
— A little boy and girl were playing in a skiff, which swung off the shore; the mother waded into the water and rescued the girl. The boy, sitting in the bottom of the skiff, with a hand on each side, went over.
1870 A lady from Chicago, said to be deranged, threw herself from Goat Island Bridge, and went over.
1871 In June three men, unacquainted with the river, hired a boat to cross, were drawn into the rapids and went over.
— In July two men in a boat went over.
1873 Friday, July 4th, a young man and woman, and a boy twelve years of age, brother of the latter, hired a boat in Chippewa, ostensibly for a sail

on the river. Not understanding the currents, they were drawn into the rapids and carried over the Horseshoe Fall. The bodies were not recovered. It was afterward ascertained that the young man had taken $500 from his father, in Ohio; had come to Chippewa to meet the young woman, who was from Toronto, to whom he was married on the day preceding their death.

1874 September 19th, a young man connected with the Mohawk Institute, at Brantford, Canada — whether as student or instructor was not known — walked deliberately into the rapids above Table Rock, and was carried over the precipice, never to be seen again.

1875 September 8th, Captain John Jones — at that time marine surveyor for a New York insurance company — jumped into the rapids below Goat Island Bridge, and went over the cliff, before the eyes of many excursionists. Ill-health was supposed to be the cause. The body was not found.

1877 March 5th, Mr. G. Homer Stone, aged twenty-four, a school-teacher, living near Geneva, N. Y., leaped into the rapids, near the upper end of Prospect Park, and was carried over the Falls. The body was not recovered.

July 1st, three men went out in a sail-boat from Connor's Island, during a high wind and very rough water. Attempting a starboard tack, in order to reach Gill Creek Island, the boat was upset, and two of them — after the three had tried in vain to right the boat, and found it difficult to keep their hold — abandoned it and tried to swim ashore; but, owing to the rough sea and their wet and heavy clothing, they were soon exhausted, and went to the bottom. The third man, divesting himself of everything except his pantaloons, determined to swim for the nearest land the down-floating boat should pass. Fortunately, a large boat, manned by three sturdy oarsmen, coming up the river, rescued him, after he had become nearly exhausted. Three days after the accident one of the bodies was found near Grass Island, above the Falls, and the other, two days later, in the Whirlpool below.

1877 October 16th, the discovery in the morning of several articles of female apparel on a flat rock, near the site of the old stone tower, and close to the brink of the Falls, led to investigation, which developed the fact that Miss Schofield, a young woman from Woodstock, in Canada, while suffering from a sudden attack of brain fever, had thrown herself into the rapids, and gone over the Horseshoe Fall. She was a skillful telegrapher, and had some local literary reputation. Her body was never recovered.

1878 April 1st, John and Patrick Reilley, brothers, started from Port Day, above the Falls, to row across to Chippewa. One of them, being under the influence of liquor, refused to row steadily and quarreled with his brother, thus preventing him from rowing. They were drawn over the Canadian side of the Horseshoe Fall about four o'clock in the afternoon. They were both skillful rowers, and well acquainted with the river, which they had crossed and recrossed many times. Their bodies were recovered several weeks later.

1878 April 6th, a young man, nineteen years of age, from Woodstock, Canada, a member of the Queen's Own, a volunteer regiment, which had attended a recent military review at Montreal, was on his return home, and crossed from Chippewa to Navy Island to visit friends who kept small boats on both sides of the river. After finishing his visit, he declined to accept the assistance of a young relative in recrossing the river, and started alone. The result was that, not understanding the force of the treacherous current, he was carried into the great rapids and went over the Horseshoe Fall. His body was found, two days afterward, below the ferry.

1879 June 21st, the names of Monsieur and Madame Rolland were registered at one of the hotels, where they spent a night, but took their meals at a restaurant kept by a Frenchman, because Monsieur R. could not, as he said, speak English. The following morning they went to the Moss Islands. While near the lower end of the outer island, so the husband claimed, madame took a cup from him to get a drink of water from the rapids, and, while his attention was diverted for a moment, he heard a splash in the water, and on looking round, saw that his wife had fallen into the rapids. She went over the Horseshoe Fall. He showed great distress and every demonstration of sorrow. Nevertheless, he left the next day for New York, after giving his address to the restaurant-keeper, who, a few days afterward, sent word to him that the body had been recovered. Monsieur R. sent thirty dollars to pay expenses of burial, and sailed for France. Those who have seen the place where, according to his story, madame fell in, are skeptical on that point.

1881 February 23d, a stranger named Doyle threw himself into the rapids from Prospect Park, and was carried over the American Fall. A body found some days after in the river below, claimed by friends to be his, was identified by a coroner's jury as that of a man named Rowell, whose body had been found some days before in the river, near the ferry, with a bullet through the head. It was never ascertained whether it was a suicide or an assassination.

— July 12th, the body of a woman was found floating below the Falls, having evidently come from the river above. Some female wearing apparel found on the shore of the rapids, below Goat Island Bridge, it was supposed belonged to the suicide.

1881 Dr. H. and Mrs. S., of good birth, education, and social position, loved not wisely but too well. Exposure was certain and near. They met at Niagara, July 14th, and went over the Falls together.

— September 5th, a man from Toronto plunged into the rapids at Table Rock, and went over. In a letter to a Toronto paper, he stated that domestic trouble was the impelling motive.

Below the Falls.

In 1841 A number of British soldiers, stationed at Drummondville, attempted to swim across the rapids at the ferry at different times. None succeeded, and two were drowned.

1842 A British soldier attempted to lower himself down the bank, opposite Barnett's Museum, in order to escape to the American shore. The rope broke, and he was killed by the fall.

1844 In August, a gentleman was washed under the great Fall, from a rock on which he had stepped, against the remonstrances of the guide. He was drowned.

1846 In August, a gentleman fell forty feet from a rock near the Cave of the Winds, and was instantly killed.

1875 August 9th, two young women and three young men, residents of the village, went through the Cave of the Winds, as they had often done

1875 before, to enjoy the exhilarating bath. One of the young women, Miss P., stepped into one of the eddying pools lying a little outside of the usual track, and one of the young men, Mr. P., thinking she might find the current stronger than she anticipated, followed her, and while seeking a sure footing for himself to guard against accident, the young lady lost her balance and fell into the current. Mr. P. endeavored to seize her bathing-dress, but not succeeding, sprang at once into the current, and both went over a ledge some eight feet high, at the foot of which Miss P. rose to her feet in an eddy, and sought support by leaning against a large rock lying adjacent to it. When Mr. P. rose to the surface he swam to her, and thinking they would be safer in an opening among smaller rocks on the opposite side of the eddy, he put his arm round her, and both made a desperate effort to reach the desired shelter. But the current proved too strong, and bore them both out into the river; Mr. P. swimming on his back, and supporting Miss P. with his right arm, while her right hand rested upon his shoulder. Suddenly they became separated. Miss P., apparently concluding that both could not be saved, disengaged herself from him, and immediately sank below the surface. Instantly her heroic friend plunged after her. A cloud of spray covered the troubled waters for a moment, and when it passed nothing could be seen of the unfortunate pair. The treacherous under-currents bore them to their doom. Both bodies were recovered a few days afterward from the Whirlpool.

1877 August 31st, Dr. Louis M. Stein registered at the International Hotel. The following day, after riding to different points on the American side of the Falls, he alighted at the upper Suspension Bridge, and inviting a young bootblack to accompany him, he started across the bridge, talking rather incoherently on the way. When near the Canadian end he stopped, took from his pocket a roll of bills, gave the boy a dollar note, and returned the others to his pocket. He then started back, and when near the center of the bridge dropped his hand-bag and shawl, seized the boy, saying with an oath, "You have got to come, too!" and attempted to climb over the railing. The boy successfully resisted, but the man got over and dropped from one of the wire stays into the river, one hundred and ninety feet below. He was probably killed instantly, and the body floated down the river, from which it was taken some ten days afterward and delivered to a son, who arrived from New York city.

— December 25th, a man from Chatauqua County, N. Y., suffering from ill-health and misfortune, jumped from the new Suspension Bridge, and was never seen again.

The narrowest escape at the Falls was that of the man who, in January, 1852, fell from the Tower Bridge into the rapids, and was caught between two rocks just on the brink of the precipice, whence he was rescued, nearly exhausted, by means of a rope.

In 1874, Mr. William McCullough, while at work painting the small bridge between the first and second Moss Islands, missed his footing and fell into the middle of the channel; he was carried down about fifty rods, and, going over a ledge into more quiet water, got on his feet and waded to a small rock projecting above the water, upon which he seated himself to collect his senses and await results. After several vain efforts to get a rope to him, Mr. Thomas Conroy, a guide, then connected with the Cave of the Winds, who had in the previous autumn conducted Professor Tyndall up to Tyndall's Rock, put on a pair of felt shoes, and, holding to an inch rope, picked his way with an alpen-stock, from a point a short distance up-stream, through favoring eddies and pools to McCullough. After a short rest, he put the rope around McCullough, under his arms, and winding the end around his own right arm, the two started shoreward. On reaching the deep water near the shore, both were taken off their feet, and, as the people pulled vigorously at the rope, their heads went under for a short distance, but they were safely landed. A contribution was taken up for Conroy's benefit, and Professor Tyndall, on hearing of the rescue, sent him a five-pound note.

In view of the fact that nearly every year persons are drawn into the rapids and carried over the Falls, a New

York journalist suggested a most extraordinary method of saving them. He proposed that a cable should be stretched across the rapids, above the Falls, strong enough to arrest boats, and to which persons in danger might cling until rescued. But this kind and ingenious person forgot that old canal-boats, rafts of logs, and large trunks of trees, with roots attached, would be troublesome things to hold at anchor. As well hope to stay an Alpine avalanche with pipe-stems.

How the Suspension Bridge was Begun.

Opposite page 137.

CHAPTER XVI.

The first Suspension Bridge—The Railway Suspension Bridge—Extraordinary vibration given to the Railway Bridge by the fall of a mass of rock—De Veaux College—The Lewiston Suspension Bridge—The Suspension Bridge at the Falls.

ON the partial completion of the Hydraulic Canal, the principal stockholders, with a number of invited guests, celebrated the event on July 4, 1857, by an excursion from Buffalo in the *Cygnet*, the first steamer that ever landed within the limits of the village of Niagara. The same route is followed during the season of navigation by tugs towing canal-boats and rafts out and in. No passenger boat, however, has been placed on the route, although the sail on the river is a charming one.

Mr. Charles Ellet, in 1840, built the first suspension bridge over the chasm. He offered a reward of five dollars to any one who would get a string across it. The next windy day all the boys in the neighborhood were kiting, and before night a youth landed his kite in Canada and received the reward. The first iron successor of the string was a small wire cable, seven-eighths of an inch in diameter. To this was suspended a wire basket in which two persons could cross the chasm. The basket was attached to an endless rope, worked by a windlass on each bank. At an entertainment given on the occasion of the completion of the bridge, the good people of the embryo village at the bridge, elated with their new acquisition, were inclined to regard their neighbors at the Falls with patronizing sympathy. One of the latter said to Mr. Ellet, "This bridge is a very clever affair, and you only need the Falls here to build up a respectable village." "Well," he replied, "give me money enough and I will put them here." He had great faith in dollar-power.

This bridge was an excellent auxiliary in the construction of the present Railway Suspension Bridge, built by Mr. John A. Roebling. It was begun in 1852, and the first locomotive crossed it in March, 1855. It is one of the most brilliant examples of modern engineering, and stands unrivaled for its grace, beauty, and strength. Seizing at once upon the natural advantages of the location, the engineer resolved to combine the tubular system with that of the suspension bridge. The carriage way was placed level with the banks of the river at the edges of the chasm. The railway track was placed eighteen feet above, on a level with the top of the secondary banks across which the two railroads were to approach it. The plan was perfect, and perfectly and faithfully executed in all its details. It is practically a skeleton tube. As the traveler passes over it in a carriage or a railway car, from the almost total absence of any vibratory motion he feels at once that he is on a safe basis, and his sense of security is complete.

One feature of the construction of the bridge may be noticed as having a bearing on the question of its durability. It is well known that when wrought-iron is exposed to long continued or oft repeated and rapid concussions, its fibers after a time become granulated, whereby its strength is greatly impaired and finally exhausted. It is also known that the effect of rhythmical or regular vibrations is more destructive than the effect of those which are inharmonious or irregular. Because of this, a body of men is never allowed to march to music across a bridge, nor is a large number of cattle ever driven across at one time, lest they should, by accident, fall into a common step and so overstrain or break down the bridge. It is the difference between a single heavy blow and an irregular succession of light ones. Hence, when harmonious, regular vibrations can be broken up, the destructive influence is greatly modified and retarded.

The bridge is supported by two large cables on each side, one pair above the other, the lower pair being nearer together horizontally than the upper pair, so that a cross section of the skeleton tube would be shaped somewhat like the keystone of an arch. Each of these large cables is ten inches in diameter, and is composed of seven smaller ones, called strands. These smaller strands are made of number nine wire, and each one contains five hundred and twenty wires. Each of these wires was boiled three several times in linseed oil, giving it an oleaginous coating of considerable thickness and great adhesive power. Each wire was carried across the river separately, from tower to tower, by a contrivance of the engineers, the chief feature of which was a light iron pulley about twenty inches in diameter, suspended on what might be called a wire cord. This apparatus was called a traveler, and curious and interesting was its performance as seen from below. It looked like a huge spider weaving an iron web.

Six of the seven strands forming each of the cables were laid around the seventh as a center, and when all were properly placed they were again saturated with oil and paint. After this, by another contrivance of the engineers, they were wound or wrapped with wire, like winding a rope cable with marlin, and thus the whole cable was made into a thoroughly compact, huge, round, iron rope. This was covered with numerous coats of paint to prevent the oxidation of the inner wires. The oleaginous coating of the wires, together with the small triangular spaces between them, would seem to reduce the destructive power of the vibrations to zero. But the vibrations are very greatly reduced and the stiffness of the structure is greatly increased by the use of a series of triangular stays, the triangle being the only geometrical figure whose angles cannot be shifted. There are sixty-four of these triangles. Their hypothenuses are formed by over-floor stays of wire rope reaching from the tops of the towers to different points in the lower floor, this latter, of course, forming their common base and the towers their altitude. The stays are fastened to the suspenders so as to form straight lines. As the towers and the floor are rigid and solid in the direction of the lines they represent, it follows that the intersections of the hypothenuses with the common base form so many stationary points in the latter. These stationary points present a powerful resistance to vibrations. The side trusses, with their system of diamond-work braces and the weight of the railway track on the upper bridge, also help to stiffen the structure. There are likewise fifty-six under stays or guys of wire rope fastened to the rocks below, designed to prevent upward and lateral vibrations. A heavy locomotive with twenty loaded cars produced a depression of the upward curvature of the track of nearly ten inches. The ordinary loads make a depression of only five inches.

In Part II., attention was directed to a point on the American side of the river, just below this bridge, where the disintegration of the shale and abrasion of the super-posed rock is strikingly exhibited. A singular phenomenon was witnessed here in 1863. A mass of rock and shale, about fifty feet long, twenty feet wide, and sixty feet deep, fell with a great crash. Directly following the fall a remarkable motion was developed in the bridge itself. A strong wave of motion passed through the whole structure from the American side to the opposite shore, and returned again to the same side.

Some twelve or fifteen mechanics, who were at work on the upper or railway track, were so alarmed that they fled with all speed to the shore. The motion imparted to the bridge was incalculably greater than, and of a different character from, any motion imparted by the crossing of the heaviest trains. The rocky mass which fell was forty rods below the bridge, and the hard floor on which it struck was more than two hundred and thirty feet beneath it. The mass itself fell about sixty feet average distance, and might have weighed five thousand tons. The extraordinary motion imparted to the bridge by the concussion must have been transmitted along the bed-rock to the anchorages on the American side, thence through the cables and the bridge across to the anchorages on the Canadian side, whence it returned to the American side.

Mr. Donald McKenzie, master carpenter and superintendent of repairs, who has been connected with the bridge constantly since its erection, and all the men under him at the time, confirm this statement, and declare it is impossible to exaggerate or describe the wave-like motion which they experienced while escaping to the shore.

Half a mile further down is De Veaux College, a noble charity endowed by the late Mr. Samuel De Veaux. He was for many years an active business man at Niagara, and by his integrity, industry, and wise enterprise accumulated a handsome fortune. His death occurred in 1852, and by his will he left nearly the whole of his estate to certain trustees to establish an institution for the care, training, and education of orphan boys. In addition to these, other pupils are received who pay a fixed price for their tuition, board, and incidentals. The institution has gained a high reputation for the thoroughness of its instruction and the excellence of its discipline. One of its sources of income is the amount received annually for admissions to the Whirlpool. Every visitor to that interesting locality will cheerfully pay the fee charged when he understands this fact.

The suspension bridge below the mountain near Lewiston, spanning the river where the water emerges from the fearful abyss through which it dashes for five miles, was built in 1856, by Mr. T. E. Serrel. The guys designed to protect it from the effect of the wind were fastened in the rocks on either side at the water's edge.

The great ice jam of 1866 tore from their fastenings, or broke off, many of these guys. Before they were replaced a terrific gale in the following autumn broke up the roadway, severed some of the suspenders, and left the structure a melancholy wreck dangling in the air.

The New Suspension Bridge, as it is called, just below the ferry at the Falls, was built in 1868. It is a light, graceful structure, standing one hundred and ninety feet above the water. Its length is twelve hundred feet, after the Brooklyn bridge the longest structure of the kind in the world, and it is the narrowest of those designed for carriage travel. To its narrowness it probably owed its safety from destruction during a fierce gale which occurred in the fall of 1869. The fastenings or dowels of several of the guys on the Canadian side were torn out, and the bridge at its center deflected downstream more than its width, so that the surface of its road-way could not be seen half its length. Then its undulations from end to end—like a stair-carpet being shaken between two persons—were frightful, and for a time it was feared that either cables or towers must give way. After the gale subsided the old guys were made fast again, new ones were added, and two two-inch steel wire cables were stretched from bank to bank, and connected with the bridge by wire stays. Wrought-iron beams were afterward placed on the bottom stringers, and channel irons on the top beams of the side trestles, all of which were strongly bolted together. These improvements added much to the strength of the whole structure, and greatly increased its ability to resist horizontal deflection.

CHAPTER XVII.

Blondin and his "ascensions"—Visit of the Prince of Wales—Grand illumination of the Falls—The steamer *Caroline*—The water-power of Niagara—Lord Dufferin and the plan of an International Park.

IN the year 1858, a short, well-rounded, fair-complexioned, light-haired Frenchman made his appearance at the Falls, and expressed a wish to put a tight-rope across the chasm below them, for the purpose of crossing on the rope and exhibiting athletic feats. He received little encouragement, but, having a Napoleonic faith in his star, he persevered, and finally obtained the necessary authority to place his rope just below the Railway Suspension Bridge. It was a well and evenly twisted rope, about two inches in diameter; and after stretching it as taught as it could be drawn, it hung in a moderate catenary curve. Commencing at the shore ends he secured stays of small rope to the large one, placing them about eight feet apart. These were made fast to the shore in such a manner that all the stays on one side of the main rope were parallel to each other from the center outward to the ends. They were made tight somewhat in the manner that tent-cords are tightened, and when the structure was complete it looked like the opposite sections of a gigantic spider-web.

At each end was a spacious inclosure, formed by a rough board fence, for the use of spectators. M. Blondin — for this was the name of the new aspirant for acrobatic honors — also made an arrangement with the superintendent of the railway bridge for its occupation during what, with a shade of irony, he called his "ascensions." Those who went within the inclosures and upon the bridge paid a certain sum. A contribution was asked of all outsiders. He selected Saturday as the day for fortnightly ascensions, and advertised his intentions very liberally. The speculation was successful and gave great satisfaction to the spectators. He exhibited a variety of rope-walking feats, balancing on the cable, hanging from it by his hands and feet, standing on his head, and lowering himself down to the surface of the water. He also carried a man across on his back, trundled over a loaded wheelbarrow, and did divers other things, and also walked over in a sack. He sprinkled in a few extras to heighten the effect, as the knowing ones declared, such as slipping astride the cable, falling across a stay-rope, or dropping something into the water. In 1860, he gave a special ascension in honor of the Prince of Wales. The Prince and his party occupied a sheltered space on the Canadian side, and Blondin walked to it from the opposite side, performing various feats on the way over. The Prince shook hands with him as he stepped into the shed, and commended his courage and nerve.

As illustrating the power of the imagination over the nerves it may be noted that, if the great spider's-web had been stretched out anywhere on a level surface, and not more than three feet above the ground, a dozen men in any large community could have been found to walk it as unconcernedly, if not as gracefully, as the famous "ascensionist." After three years of successful labor at Niagara, he sought other air-spaces.

The most notable occurrence, however, which emphasized the visit of the Prince of Wales in that year was the illumination of the Falls late in the evening of a moonless night. On the banks above and all about on the rocks below, on the lower side of the road down the Canadian bank, and along the water's edge, were placed numerous colored and white calcium, volcanic, and torpedo lights. At a signal they were set aflame all at once. At the same time rockets and wheels and flying artillery were set off in great abundance. The shores were crowded with spectators, and the scene was a most remarkable one. The steady, lurid light below and the intermittent flashes and explosions overhead, the seething, hissing volumes of flame and smoke rolling up from the deep abyss, the ghostly appearance of the descending stream, the ghastly swift current of white foam, the weird appearance of the cloud of spray with a faint and fantastic illumination at its base, which faded out in the dim light of the stars as it ascended, the peculiarly deep but muffled and solemn monotone of the falling water, the livid hue imparted to the faces of the quiet but deeply interested spectators, all made the scene memorable and impressive. When the Marquis of Lorne and the Princess Louise visited the Falls in January, 1879, they saw them illuminated by electricity, the light having the illuminating power of 32,000 candles.

In December, 1837, the steamer *Caroline* came down from Buffalo to aid, it was said, the so-called Patriots, then engaged in an insurrection against the Canadian Government. A motley collection of adventurers on Navy Island constituted the disturbing, not to say attacking, force. At Chippewa was stationed a body of Canadian militia, under the command of Colonel—afterward Sir—Allan McNabb, who had the good fortune to win his spurs in a single almost bloodless campaign. By his direction a boat expedition was sent to attack the *Caroline*, as she lay at the old Schlosser dock. In the mêlée one American was killed. The steamer was set on fire, and her fastenings must have been burnt away, as also a part of her upper works, since the writer, ten years later, while returning from a fishing expedition, discovered her smoke-pipe lying at the bottom of the river, in a quiet basin not thirty rods below the dock. A cat-fish of moderate dimensions appeared to be keeping house in it, and, with his head barely projecting from one end, was serenely watching the current for whatever game it might bring to his iron parlor. After the new bridges were built connecting the Three Sisters with Goat Island, the guides and drivers, in their desire to enhance the interest of the scene, astonished travelers by informing them that it was the boiler of the *Caroline* which caused the extraordinary elevation of the water which we have before referred to as the Leaping Rock.

Nine miles from the Falls is the Tuscarora Reservation of four thousand acres. On this there are about three hundred and fifty Indians, mostly half-breeds, engaged in agricultural pursuits, which supply a portion of their necessities. The Indian women who are seen at the Falls in the summer season working and vending different articles of bead-work belong to this community. The Tuscaroras have not been more fortunate than others of their race in bargaining with their white brothers, and their lands are now stripped of the fine oak timber and valuable wood which stood upon it a few years since, and which was sold in large quantities at small prices.

As a compensation for this system of robbery we maintained a Christian missionary among them for a few years, and we boast that they are all Protestants. The resident missionary, a very worthy man, but a rather prosy preacher, always addressed his dusky audience in the English language, his thoughts being conveyed to them by an interpreter. For many years the interpreter was a native Tuscarora, a fine specimen of his race, six feet tall, with a tawny complexion, dark, flashing eyes, and a musical voice. It was interesting to note his manner while acting as interpreter for different clergymen. When interpreting the pious but humdrum utterances of the passionless missionary, he stood at the right side of the preacher, with his left elbow resting on one end of the modest pulpit, and delivered himself with an air that seemed to say, "It does not amount to much, but I give it to you as it is." But the change was magical when, as sometimes happened during the summer season, some eloquent preacher addressed the congregation. The natural courtesy of the interpreter led him, instead of putting his elbow on

Indian Women Selling Bead-work.

the pulpit, to stand a little to the rear of the strange preacher, respectfully waiting for his words. As the priest warmed into his subject the interpreter caught his spirit, straightened his fine figure to its full height, advanced to a line with the speaker, and as the theme was developed and the orator grew more and more eloquent, the excitement became contagious ; the Indian entered fully into its spirit, his face glowed with animation, his eyes shone with a warmer light, his long arms were stretched forth, and with gestures energetic or subdued, but always graceful, and the varied inflections of his voice in harmony with the theme, he followed the discourse to the end. His audience, too, would become thoroughly aroused, and a little more animation would be infused into the plaintive tones of the closing hymn.

One of the future attractions of Niagara, to sportsmen at least, may be the catching of California trout, twenty thousand of the fry having been put into the rapids by the writer in June, 1881.

Concerning the manufactories, shops, rubbish, and litter along the race near the brink of the American Falls, which appear so uncouth and inharmonious, and which are noticed by strangers as being a desecration of the scene, it is only just to remark that the utilization of the water-power here, in the easiest and most economical manner, was one of the imperative necessities of the early settlement of the country. For many years a large territory, lying on both sides of the river, was dependent upon the manufacturing, repairing, and milling facilities of this place. For furnishing these in those days, water-power was the only agent. And the name—Manchester—given to the place by its early settlers only foreshadowed their hope that it would one day rival its great English namesake.

There are fewer manufactories on the old race-ways now than there were forty years ago, but many new ones have been located on the hydraulic canal that has been excavated at great expense, which leaves the river a mile above the Falls, and empties into the chasm half a mile below. The three years of unusual drought in the northern half of the United States, from 1876 forward, demonstrated how little dependence can be placed during the summer season on the ordinary water-powers of that region, and the attention of manufacturers has been newly drawn to Niagara.

The early dream of growth in population and wealth at Niagara seems likely to be realized. Already extensive milling and manufacturing establishments have been put in operation, and others are in contemplation. When it is considered that engineers estimate the sum-total of all the water-power in the northern portion of the United States at less than 500,000 horse-power, and that, according to data furnished by the United States Lake Survey Bureau, the water-power of Niagara is equal to 1,500,000 horse-power, we can form some idea of the vastness of the force which awaits the enterprise of American manufacturers.

"I understand, Mr. President," said Daniel Webster, in a speech prefacing a toast complimentary to the citizens of Rochester for their generous hospitality at the

New York State Fair in 1844, " that the Genesee River has a fall of 250 feet within the limits of the city of Rochester. Sir, if the Thames had a fall of 250 feet within the limits of the city of London, London would not be a town — it would be a-l-l t-h-e w-o-r-l-d !" and as he deliberately stretched out his great arms, and expanded his broad chest, while slowly pronouncing the last three words, one could almost see London gradually enlarging its ample borders in all directions. When the 1,500,000 horse-power of Niagara is utilized for the economic wants of men, Niagara will not be a town — it will be a large part of all the world.

On the 25th of September, 1878, in an after-luncheon speech before the Ontario Society of Artists at Toronto, Lord Dufferin, Governor-General of Canada, first publicly suggested the idea of creating an International Park from lands to be taken from both sides of the river adjacent to and including the Falls. He stated that he had conferred with Governor Robinson of New York upon the subject, and that the project was cordially approved by him. Governor Robinson, in his annual message the following winter, commended the project to the consideration of the Legislature, by whom a commission of distinguished gentlemen was appointed to investigate the subject and report thereon. After a full examination this commission reported warmly in favor of the plan, and their recommendation was cordially indorsed by a great many prominent citizens residing in different sections of the country. The press, too, was almost unanimously for it. A majority of the members of the Legislature to whom the report was made would have passed a bill for the further prosecution of the scheme, but, unfortunately, it was ascertained that any bill they might pass for this purpose would be vetoed for economical reasons. It is hoped that better counsels may ultimately prevail, and the plan be perfected. Nothing else can save Niagara from total desecration and disgrace. The fact that there is not a square foot of land in the United States from which an untaxed view of the great cataract can be obtained is a disgrace to the State, the nation, and the civilization of the age.

CHAPTER XVIII.

Poetry in the Table Rock albums — Poems by Colonel Porter, Willis G. Clark, Lord Morpeth, José Maria Heredia, A. S. Ridgely, Mrs. Sigourney, and J. G. C. Brainard.

BEFORE the last fall of Table Rock, there stood upon it for many years a comfortable summer-house, where people could take refuge from the spray, look at the Falls, partake of luncheon, and procure guides and dresses to go under the sheet. In the sitting-room was a large round table, on which were placed a number of albums, as they were called. In these visitors could write whatever thoughts or sentiments might be suggested by the scene. With the grand reality before them but few persons attempted anything serious, by far the greater number adopting the facetious vein. It was emphatically light literature. One or two collections of it have been published, furnishing the reader with only a modicum of sense to an intolerable quantity of nonsense.

The following specimens are better than the average:

"To view Niagara Falls, one day,
A Parson and a Tailor took their way.
The Parson cried, while rapt in wonder
And list'ning to the cataract's thunder:
'Lord! how thy works amaze our eyes,
And fill our hearts with vast surprise!'
The Tailor merely made this note:
'Lord! what a place to sponge a coat!'"

"THOUGHTS ON VISITING NIAGARA.

" I wonder how long you've been a roarin'
  At this infernal rate:
 I wonder if all you've been a pourin'
  Could be ciphered on a slate.

" I wonder how such a thund'rin' sounded
  When all New York was woods;
 I suppose some Indians have been drownded
  When rains have raised your floods.

" I wonder if wild stags and buffaloes
  Hav'nt stood where now I stand;
 Well, 'spose — bein' scared at first — they stub'd their toes,
  I wonder where they'd land!

" I wonder if the rainbow's been a shinin'
  Since sunrise at creation;
 And this waterfall been underminin'
  With constant spatteration!

" That Moses never mentioned ye, I've wonder'd,
  While other things describin';
 My conscience! how loud you must have thunder'd
  While the deluge was subsidin'!

" My thoughts are strange, magnificent, and deep
  While I look down on thee.
 Oh! what a splendid place for washing sheep
  Niagara would be!

" And oh! what a tremendous water power
  Is wasted o'er its edge!
 One man might furnish all the world with flour
  With a single privilege.

" I wonder how many times the lakes have all
  Been emptied over here?

Why Clinton didn't feed the Grand Canawl
From hence, I think is queer."

The most graceful verses on Niagara ever written by a resident are the following by the late Colonel Porter, who was an artist both with the pencil and the pen. They were written for a young relative in playful explanation of a sketch he had drawn at the top of a page in her album, representing the Falls in the distance, and an Indian chief and two Europeans in the foreground:

" An Artist, underneath his sign (a masterpiece, of course)
Had written, to prevent mistakes, 'This represents a horse':
So ere I send my Album Sketch, lest connoisseurs should err,
I think it well my Pen should be my Art's interpreter.

" A chieftain of the Iroquois, clad in a bison's skin,
Had led two travelers through the wood, La Salle and Hennepin.
He points, and there they, standing, gaze upon the ceaseless flow
Of waters falling as they fell two hundred years ago.

" Those three are gone, and little heed our worldly gain or loss —
The Chief, the Soldier of the Sword, the Soldier of the Cross.
One died in battle, one in bed, and one by secret foe;
But the waters fall as once they fell two hundred years ago.

" Ah, me! what myriads of men, since then, have come and gone;
What states have risen and decayed, what prizes lost and won;
What varied tricks the juggler, Time, has played with all below:
But the waters fall as once they fell two hundred years ago.

" What troops of tourists have encamped upon the river's brink;
What poets shed from countless quills Niagaras of ink;
What artist armies tried to fix the evanescent bow
Of the waters falling as they fell two hundred years ago.

"And stately inns feed scores of guests from well replenished larder,
And hackmen drive their horses hard, but drive a bargain harder;
And screaming locomotives rush in anger to and fro:
But the waters fall as once they fell two hundred years ago.

"And brides of every age and clime frequent the island's bower,
And gaze from off the stone-built perch — hence called the Bridal Tower —
And many a lunar belle goes forth to meet a lunar beau,
By the waters falling as they fell two hundred years ago.

"And bridges bind thy breast, O stream! and buzzing mill-wheels turn,
To show, like Samson, thou art forced thy daily bread to earn:
And steamers splash thy milk-white waves, exulting as they go,
But the waters fall as once they fell two hundred years ago.

"Thy banks no longer are the same that early travelers found them,
But break and crumble now and then like other banks around them;
And on their verge our life sweeps on — alternate joy and woe;
But the waters fall as once they fell two hundred years ago.

"Thus phantoms of a by-gone age have melted like the spray,
And in our turn we too shall pass, the phantoms of to-day:
But the armies of the coming time shall watch the ceaseless flow
Of waters falling as they fell two hundred years ago."

On turning to the more serious poems that have been written on the theme, the reader naturally experiences a feeling of disappointment that a scene which has filled and charmed so many eyes should have found so few interpreters. Only those who see Niagara know how fast the tongue is bound when the thought struggles most for utterance. One who seems to have experienced this feeling thus expresses it:

> " I came to see;
> I thought to write;
> I am but —— dumb."

The late Mr. Willis G. Clark thus expands the same sentiment:

> " Here speaks the voice of God—let man be dumb,
> Nor with his vain aspiring hither come.
> That voice impels the hollow-sounding floods,
> And like a Presence fills the distant woods.
> These groaning rocks the Almighty's finger piled;
> For ages here his painted bow has smiled,
> Mocking the changes and the chance of time—
> Eternal, beautiful, serene, sublime!"

The following from the Table Rock Album was written by the late Lord Morpeth:

NIAGARA FALLS.— BY LORD MORPETH.

> " There's nothing great or bright, thou glorious Fall!
> Thou mayest not to the fancy's sense recall.
> The thunder-riven cloud, the lightning's leap,
> The stirring of the chambers of the deep;
> Earth's emerald green and many tinted dyes,
> The fleecy whiteness of the upper skies;
> The tread of armies thickening as they come.
> The boom of cannon and the beat of drum;

The brow of beauty and the form of grace,
The passion and the prowess of our race;
The song of Homer in its loftiest hour,
The unresisted sweep of human power;
Britannia's trident on the azure sea,
America's young shout of Liberty!
Oh! may the waves which madden in thy deep
There spend their rage nor climb the encircling steep;
And till the conflict of thy surges cease
The nations on thy banks repose in peace."

The extracts below are from a poem written after a visit to the Falls by José Maria Heredia, and translated from the Spanish by William Cullen Bryant:

"NIAGARA.

" Tremendous torrent! for an instant hush
The terrors of thy voice, and cast aside
Those wide involving shadows, that my eyes
May see the fearful beauty of thy face!

. . . . . .

" Thou flowest on in quiet, till thy waves
Grow broken 'midst the rocks; thy current then
Shoots onward like the irresistible course
Of destiny. Ah, terribly they rage,—
The hoarse and rapid whirlpools there! My brain
Grows wild, my senses wander, as I gaze
Upon the hurrying waters; and my sight
Vainly would follow, as toward the verge
Sweeps the wide torrent. Waves innumerable
Meet there and madden,—waves innumerable
Urge on and overtake the waves before,
And disappear in thunder and in foam.

" They reach, they leap the barrier,— the abyss
Swallows insatiable the sinking waves.
A thousand rainbows arch them, and woods
Are deafened with the roar. The violent shock
Shatters to vapor the descending sheets.
A cloudy whirlwind fills the gulf, and heaves
The mighty pyramid of circling mist
To heaven.   *   *   *   *
What seeks my restless eye? Why are not here,
About the jaws of this abyss, the palms,—
Ah, the delicious palms,— that on the plains
Of my own native Cuba spring and spread
Their thickly foliaged summits to the sun,
And, in the breathings of the ocean air
Wave soft beneath the heaven's unspotted blue?

" But no, Niagara,— thy forest pines
Are fitter coronal for thee. The palm,
The effeminate myrtle and pale rose may grow
In gardens and give out their fragrance there,
Unmanning him who breathes it. Thine it is
To do a nobler office. Generous minds
Behold thee, and are moved and learn to rise
Above earth's frivolous pleasures ; they partake
Thy grandeur at the utterance of thy name.

           *   *   *   *   *   *

" Dread torrent, that with wonder and with fear
Dost overwhelm the soul of him who looks
Upon thee, and dost bear it from itself,—
Whence hast thou thy beginning? Who supplies,
Age after age, thy unexhausted springs?
What power hath ordered that, when all thy weight
Descends into the deep, the swollen waves
Rise not and roll to overwhelm the earth?

> "The Lord hath opened his omnipotent hand,
> Covered thy face with clouds and given his voice
> To thy down-rushing waters: he hath girt
> Thy terrible forehead with his radiant bow.
> I see thy never-resting waters run,
> And I bethink me how the tide of time
> Sweeps to eternity."

The lyric from which the following extracts are taken was written by Mr. A. S. Ridgely, of Baltimore, Md.:

> " Man lays his scepter on the ocean waste,
> His footprints stiffen in the Alpine snows,
> But only God moves visibly in thee,
> O King of Floods! that with resistless fate
> Down plungest in thy mighty width and depth.
> * * * Amazement, terror, fill,
> Impress and overcome the gazer's soul.
> Man's schemes and dreams and petty littleness
> Lie open and revealed. Himself far less—
> Kneeling before thy great confessional—
> Than are the bubbles of the passing tides.
> Words may not picture thee, nor pencil paint
> Thy might of waters, volumed vast and deep;
> Thy many-toned and all-pervading voice;
> Thy wood-crown'd Isle, fast anchor'd on the brink
> Of the dread precipice; thy double stream,
> Divided, yet in beauty unimpaired;
> Thy wat'ry caverns and thy crystal walls;
> Thy crest of sunlight and thy depths of shade,
> Boiling and seething like a Phlegethon
> Amid the wind-swept and convolving spray,
> Steady as Faith and beautiful as Hope.
> There, of beam and cloud the fair creation,
> The rainbow arches its ethereal hues.
> From flint and granite in compacture strong,

Not with steel thrice harden'd—but with the wave
Soft and translucent—did the new-born Time
Chisel thy altars. Here hast thou ever poured
Earth's grand libation to Eternity;
Thy misty incense rising unto God—
The God that was and is and is to be."

Mrs. Sigourney wrote the following poem, it is said, during a visit to Table Rock:

"APOSTROPHE TO NIAGARA.

"Flow on, forever, in thy glorious robe
Of terror and of beauty. God has set
His rainbow on thy forehead, and the clouds
Mantled around thy feet. And He doth give
Thy voice of thunder power to speak of Him
Eternally, bidding the lip of man
Keep silence, and upon thy rocky altar pour
Incense of awe-struck praise.
     And who can dare
To lift the insect trump of earthly hope,
Or love, or sorrow, 'mid the peal sublime
Of thy tremendous hymn! Even ocean shrinks
Back from thy brotherhood, and his wild waves
Retire abashed; for he doth sometimes seem
To sleep like a spent laborer, and recall
His wearied billows from their vieing play,
And lull them to a cradle calm; but thou,
With everlasting, undecaying tide
Dost rest not night nor day.
     The morning stars,
When first they sang o'er young creation's birth,
Heard thy deep anthem; and those wrecking fires
That wait the archangel's signal, to dissolve
The solid earth, shall find Jehovah's name

Graven, as with a thousand spears
On thine unfathomed page. Each leafy bough
That lifts itself within thy proud domain
Doth gather greenness from thy living spray,
And tremble at the baptism. Lo! yon birds
Do venture boldly near, bathing their wings
Amid thy foam and mist. 'Tis meet for them
To touch thy garment here, or lightly stir
The snowy leaflets of this vapor wreath,
Who sport unharmed on the fleecy cloud,
And listen to the echoing gate of heaven
Without reproof. But as for us, it seems
Scarce lawful with our broken tones to speak
Familiarly of thee. Methinks, to tint
Thy glorious features with our pencil's point,
Or woo thee with the tablet of a song,
Were profanation.
     Thou dost make the soul
A wondering witness of thy majesty;
And while it rushes with delirious joy
To tread thy vestibule, dost chain its step,
And check its rapture, with the humbling view
Of its own nothingness, bidding it stand
In the dread presence of the Invisible,
As if to answer to its God through thee."

The following lines were written by the late John G. C. Brainard, who never saw the Falls. They were dashed off at a single short sitting, for the head of the literary column of the *Connecticut Mirror*, of Hartford, which he then edited:

"THE FALLS OF NIAGARA.

" The thoughts are strange that crowd into my brain
While I look upward to thee. It would seem

As if God pour'd thee from his 'hollow hand'
And hung his bow upon thine awful front,
And spoke in that loud voice which seem'd to him
Who dwelt in Patmos for his Saviour's sake,
'The sound of many waters,' and had bade
Thy flood to chronicle the ages back,
And notch his cen'tries in the eternal rocks.

" Deep calleth unto deep. And what are we
That hear the question of that voice sublime?
Oh! what are all the notes that ever rung
From War's vain trumpet by thy thundering side!
Yea, what is all the riot man can make
In his short life to thy unceasing roar!
And yet, bold babbler, what art thou to HIM
Who drown'd a world and heap'd the waters far
Above its loftiest mountains?—a light wave
That breaks and whispers of its Maker's might."

PART IV.

OTHER FAMOUS CATARACTS OF THE WORLD.

CHAPTER XIX.

Yosemite — Vernal — Nevada — Yellowstone — Shoshone — St. Maurice — Montmorency.

FOR the purpose of comparison it may be interesting to note other cataracts in the United States, and in other parts of the world, and also some of the remarkable rapids, which may be successors to what were once perpendicular falls. For descriptions of those in foreign countries we are chiefly indebted to the geographical gazetteers and the journals of Humboldt, Livingstone, Bohle, and Stanley; for information regarding the cataracts of Norway we are indebted to Murray's "Norway, Denmark and Sweden."

In the United States, after Niagara, the first to claim our attention are the Falls of the Yosemite, so graphically and scientifically made known to us in the second volume of Professor J. D. Whitney's Geological Report for California.

Yosemite Falls.

Before describing them it is necessary to note the physical features of the region in which they are placed. The valley of the Yosemite forms a portion of the bed of the Merced River, which flows through it and passes from it by a wild, deep cañon into the San Joaquin. It is about eight miles long and from half a mile to a mile wide, with a sharp bend to the west, about two miles from its upper end. To this place the Merced and two tributaries, called the North and South Forks, have come through the most rugged cañons, falling nearly two thousand feet in the space of two miles.

Near the southerly end of the valley is the remarkable rock El Capitan, an almost vertical cliff 3,600 feet high, and one of the grandest objects in the valley. Just above this is the imposing pile called the Cathedral Rocks, and behind these, connected with them, two slender and beautiful granite columns called the Cathedral Spires.

Two miles above, on the opposite side, is the row of summits, rising like steps one above another, named the Three Brothers. On the other side, in the angle of the valley, stands Sentinel Rock, so called from its fancied resemblance to a watch-tower. Three-fourths of a mile in a southerly direction from this is the Sentinel Dome, more than four thousand feet high and affording from its summit a most magnificent view. Following up the North Fork, just at the entrance of the cañon, rises the Half Dome, the grandest and loftiest in the Yosemite Valley, an inaccessible crest of granite, having an elevation—according to Prof. Brewer—of 6,000 feet. On the opposite side of the same cañon stands the North Dome, another of those rounded masses of granite so characteristic of the sierras. Appearing as a buttress to this is Washington's Column, and below this the Royal Arches, an immense arched cavity, formed by the giving way and sliding down of portions of the rock, and presenting, in the upper part, a vaulted appearance.

In the angle formed by the Merced with the South Fork is the symmetrical and beautiful North Dome. This valley is the most remarkable basin thus far found in the world, and in view of its gigantic and impressive scenery we cannot but marvel at its size — a mere cup or trough in the midst of one of the sublimest of geological formations. This tiny strip of wonder-land is, as we have seen, only eight miles long and less than three-quarters of a mile average width.

Beginning at the south-westerly end of the valley we first reach, in ascending it, the Bridal Veil, formed by one of the torrents that feed the Merced River. It is 1,000 feet in height, the body of water not being large, but sufficient to produce the most picturesque effect. As it is swayed backward and forward by the force of the wind, it seems to flutter like a white veil.

Near the head of the valley, where it turns sharply toward the west, we have before us the Yosemite Fall. "From the edge of the cliff to the bottom of the valley the perpendicular distance is, in round numbers, 2,550 feet. The fall is not one perpendicular sheet. There is first a vertical descent of 1,500 feet, when the water strikes on

Opposite page 166.   Bridal Veil Fall.

what seems to be a projecting ledge, but which is in reality a shelf or recess about a third of a mile back from the front of the lower portion of the cliff. Across this shelf the water rushes downward in a foaming torrent on a slope, equal to a perpendicular height of 626 feet, when it makes a final plunge of about 400 feet on to a low talus of rock at the foot of the precipice. As these various falls are in one vertical plane, the effect of the whole from the opposite side of the valley is nearly as grand, and perhaps even more picturesque, than it would be if the descent was made in one sheet from the top to the bottom. The mass of water in the 1,500 feet fall is too great to allow of its being entirely broken up into spray, but it widens very much as it descends, and as the sheet vibrates backward and forward with the varying pressure of the wind, which acts with immense force on this long column of water, the effect is indescribably grand."

The first fall in the cañon of the Merced is the Vernal, "a simple perpendicular sheet 475 feet high, the rock behind it being a perfectly square-cut mass of granite. Ascending to the summit of the Vernal Fall by a series of ladders, and passing a succession of rapids and cascades of great beauty, we come to the last great fall of the Merced—the Nevada, which has a descent of 639 feet, and near its summit has a peculiar twist caused by the mass of water falling on a projecting ledge which throws it off to one side, adding greatly to the picturesque effect. It must be ranked as one of the finest cataracts in the world, taking into consideration its height, the volume and purity of the water, and the whole character of the scenery which surrounds it."

The fall from end to end of the valley proper is about fifty feet. "Its smooth and brilliant color, diversified as it is with groves of trees and carpeted with showy flowers, offers the most wonderful contrast to the towering masses of neutral and light purple-tinted rocks by which it is surrounded. Its elevation above the sea is estimated at 4,060 feet, and the cliffs and domes about it from 3,000 to 5,000 feet higher." It is a source of great satisfaction to the lover of nature that this famous and favored territory, so studded with grandeur and fretted with beauty, has wisely been set apart by Governmental authority to minister to the higher needs and better instincts of man.

The valley of the Yellowstone east of the Rocky Mountains in the north, like that of the Yosemite west of the sierras of the Pacific slope, is another wonderland, presenting a bewildering variety of land and water formations which, in turn, awe, charm, fascinate, or amuse, but always astonish, the beholder.

Among the most interesting objects in the Yellowstone Valley are the upper and lower falls of the Yellowstone River. "No language," says Professor Hayden, "can do justice to the wonderful grandeur and beauty of these scenes, and it is only through the eye that the mind can gather anything like an adequate conception of them. The two falls are not more than a fourth of a mile apart. Above the upper fall the Yellowstone flows through a grassy, meadow-like valley with a calm, steady current, giving no warning until very near the fall that it is about to rush over a precipice 140 feet high, and then, within a quarter of a mile, again leap down a distance of 350 feet. After the waters roll over the upper descent they flow with great rapidity along the upper flat, rocky bottom which spreads out to near double the width above the falls, and continues thus until near the fall, when the channel again contracts and the waters seem, as it were, to gather into a compact mass and plunge over the descent of 350 feet in detached drops of foam as white as snow."

On the Snake or Lewis River, the largest tributary of the Columbia River, are three falls, the greatest of which is the Shoshone in Idaho, where the river, with a width of six hundred yards, is said to be of so great a depth that it discharges nearly as much water as the Niagara, over a precipice about two hundred feet high. This grand fall is situated in the midst of magnificent scenery, and is surrounded by a fertile country.

Another lesser Niagara is found in the north-east, in the river St. Maurice, the largest tributary of the St. Lawrence, which falls into it from the north below Three Rivers and about twenty-two miles above its mouth. The fall—the Shawenegan—is the same height as Niagara, and while the width and depth of the river are not given, the volume of water pouring over the precipice is said to be forty thousand feet per second, a supply sufficient to produce a grand and impressive cataract.

Eight miles below Quebec the river Montmorency discharges directly into the St. Lawrence, over a cliff two hundred and fifty feet high, with a width of one hundred and fifty feet. The falling foam-flecked sheet presents a beautiful and picturesque appearance. It is unique as being the only known instance in which a tributary falls perpendicularly into the main stream.

Opposite page 174.   Nevada Falls.

CHAPTER XX.

Tequendama — Kaieteur — Paulo Affonso — Keel-fos — Riunkan-fos — Sarp-fos — Staubbach — Zambesi or Victoria — Murchison — Cavery — Schaff-hausen.

IN South America is the remarkable fall of Tequendama, on the river Bogota, which, at this point, is only one hundred and forty feet wide, and is divided into numerous narrow and deep channels which finally unite in two of nearly the same width, and make a perpendicular plunge of six hundred and fifty feet to the plain below. "The cataract," says Humboldt, "forms an assemblage of everything that is sublimely picturesque in beautiful scenery. It is not one of the highest falls, but there scarcely exists a cataract which, from so lofty a height, precipitates so voluminous a mass of water. The body, when it first parts from its bed, forms a broad arch of glassy appearance; a little lower down it assumes a fleecy form, and ultimately, in its progress, it shoots forth in millions of smaller masses, which chase each other like sky-rockets. The attending noises are quite astounding, and dense clouds of vapor soar upward, presenting beautiful rainbows in their ascent. What gives a remarkable appearance to the scene is the great difference in the vegetation surrounding different parts of it." At the summit the traveler "finds himself surrounded, not only with begonias and the yellow bark tree (Sandal), but with oaks, elms, and other plants, the growth of which recall to mind the vegetation of Europe, when suddenly he discovers, as from a terrace and at his feet, a country producing the palm, the banana, and the sugar-cane. The cause of the difference is not ascertained, the difference of altitude — one hundred and seventy-five metres — not being sufficient to exert much influence on the atmosphere."

Another and grander South American fall, of comparatively recent discovery, is the Kaieteur, so called, in the river Potaro, a large affluent of the Essequibo, the largest river in British Guiana. The volume of water is greater than that in the Bogota, and falls in a single column of dazzling whiteness seven hundred and forty feet into a vast basin below. The ascending cloud of spray, the solemn monotone of the descending flood, the extreme wildness of the primitive forest, and the luxuriant and abundant growth of tropical vines and shrubs, and their gorgeous colors, make the scene impressive.

"There is in Brazil," says Elisée Reclus, "not far from Bahia, the wonderful cataract of San Francisco, known by the name of Paulo Affonso. At the foot of a long slope over which it glides in rapids, the river, one of the most considerable of the South American continent, whirls round and round as it enters a kind of funnel-shaped cavity, roughened with rocks, and suddenly contracting its width, dashes against three rocky masses reared up like towers at the edge of the abyss ; then dividing into four vast columns of water, it plunges down into a gulf two hundred and forty-six feet in depth. The principal column,

Lower Falls of the Yellowstone.

being confined in a perpendicular passage, is scarcely sixty-six feet in width, but it must be of an enormous thickness (depth), as it forms almost the whole body of the river. Half way up, the channel which contains it bends to the left, and the falling mass, changing its direction, passes under a vertical column of water, which penetrates through it from one side to the other, and breaking it up into a chaos of surges, converts it into a sea of foam. Sometimes the white, misty vapor may be seen, and the thunder of the water may be heard at a distance of more than fifteen miles." The spray and roar of Niagara are often seen and heard at Toronto, forty miles away, across Lake Ontario.

In Norway is found the highest perpendicular fall in the world that is constantly supplied with water. It is the Keel-fos, formed by a mountain stream that falls two thousand feet into the Navöens Fjord near Gudhaven, but the water becomes a mere billowy bank of mist before it reaches the bottom.

The Riunkan-fos is another Norwegian cataract in the outlet of Lake Mjösvard, which pours through a wild, rock-studded slope until it reaches a precipice, on the brink of which it is divided by a huge mass of rock into two channels. Thence it falls eight hundred and eighty feet into a dark basin at its foot, from which water-rockets and sharp jets of foam shoot up and out in all directions. The intense whiteness of the fleecy column is indescribable.

A still more famous Norwegian cararact is the Sarp-fos in the Stor-Elven, formed by the junction of the

Lougen and Glommen, the largest of the Norwegian rivers. Like the Riunkan-fos the stream is greatly contracted in a rocky gorge, and at the edge of the cliff is divided into two channels which, however, soon unite in a fall of one hundred feet upon huge masses of rock, through and over which it rushes tumultuously for a short distance, and then flows quietly into the sea. The volume of water is unusully large for a purely mountain river, being in the gorge at the top of the fall one hundred and fifty feet wide and forty feet deep. The massive and intensely white column contrasted with the dark green foliage of the solemn pines, and the darker rocks about it, and the deep blue water into which it falls, produce a vivid impression on the mind of the beholder. The Stor-Elven here presents the curious phenomenon of a stream changing, not from a perpendicular fall to a rapid, but the reverse, from a rapid to a perpendicular fall. A great portion of the right bank of the river at the fall, and for a considerable distance below, is chiefly composed of a stiff blue clay, and the river once flowed past Sarpsborg, a mile below, in a succession of magnificent rapids. At that time a superb mansion with numerous out-buildings stood at the termination of the rapids. On the 5th of February, 1702, the mansion, together with everything in and about it, sunk into an abyss six hundred feet deep, and was entirely buried beneath the water. The walls of the house were of unusual strength and thickness, with several high towers, but the whole was buried out of sight. Fourteen persons and two hundred head of cattle were also engulfed. The catastrophe was caused by the washing

Upper Falls of the Yellowstone.

out of the blue clay, and the undermining of the bank, which then toppled over into the watery chasm.

In Switzerland is the Staubbach—dust-stream—a well known fall in the canton of Berne. It has a sheer descent of nearly nine hundred feet, in which the water is converted into spray that is easily moved by the wind, thus giving it a singularly beautiful resemblance to a white curtain floating in the air.

In South Africa, Livingstone has made the public acquainted with that extraordinary hiatus in the crust of the earth in which the great river Zambesi is swallowed up. A stream more than a thousand yards wide, dotted with islands, flowing between fertile banks clothed with the luxuriant and gorgeous vegetation of the tropics, without the least preliminary break or rapid, suddenly drops into a dark chasm of unknown depth, which, repeatedly doubling on itself, pursues its tortuous course some forty miles through the hills before emerging again into the sunlight. "From Kalai," says Livingstone, "after some twenty minutes' sail we came in sight of the columns of vapor appropriately called smoke. * * * Five columns now arose, and, bending in the direction of the wind, they seemed placed against a low ridge covered with trees. The tops of the columns at this distance (six miles) appeared to mingle with the clouds. The whole scene was extremely beautiful." At the brink of the chasm he found the river divided into two channels of unequal width by a large island called the "Garden," on account of its rich vegetation. "Creeping with awe to the verge I peered down into a large rent which had been made from bank to bank of the broad Zambesi, and saw that a stream a thousand yards broad leaped down a hundred feet and then became suddenly compressed into a space of fifteen or twenty yards. In looking down into this fissure on the right of the island one sees nothing but a dense, white cloud. From this cloud rushed up a great jet of vapor exactly like steam, and it mounted two hundred or three hundred feet high; then, condensing, it changed its hue into that of dark smoke, and came back in a constant shower. This shower fell chiefly on the opposite side of the fissure, and a few yards back from the top there stands a straight hedge of evergreen trees, whose leaves are always wet. From their roots a number of little rills run back into the gulf, but as they flow down the steep wall the column of vapor in its ascent licks them up clean off the rock, and away they mount again. They are constantly running down, but never reach the bottom."

In Northern Africa the Murchison Falls in the White Nile, between lakes Victoria N'yanzi and Albert N'yanzi, were discovered by Sir Samuel Baker, and are described by him. "Upon rounding the corner a magnificent sight burst suddenly upon us. On either side of the river were beautifully wooded cliffs rising abruptly to a height of about three hundred feet; rocks were jutting out from the intensely green foliage, and, rushing through a gap that cleft the river exactly before us, the river itself, contracted from a grand stream, was pent up in a narrow gorge scarcely fifty yards in width; roaring furiously through the rock-bound pass, it plunged in one leap of

Opposite page 176.   The Staubbach—Switzerland.

about one hundred and twenty feet perpendicularly into a dark abyss below. The fall of water was snow-white, which had a superb effect, as it contrasted with the dark cliffs that walled the river, while graceful palms of the tropics and wild plantains perfected the beauty of the view."

A writer in Hamilton's "East Indian Gazetteer" gives us an account of the cataract of Gungani Chuki in the northern branch of the river Cavery. " Much the larger stream is broken by projecting masses of rock into one cataract of prodigious volume and three or four smaller torrents. The first plunges into the river below from a height variously estimated at from one hundred to one hundred and fifty feet, while the others, impeded in their course by intervening rocks, work their way with many fantastic evolutions to a distance about two hundred feet from the base of the precipice, where they all unite to make a single final plunge, while the other branch of the river precipitates itself in two columns from a cliff of the same height, and standing nearly at right angles with the main fall. ' The surrounding scenery is wild in the extreme, and the whole presents a very imposing spectacle.

" A second cataract is formed by the southern arm of the Cavery about a mile below. The channel here spreads out into a magnificent expanse, which is divided into no less than ten distinct torrents, which fall with infinite variety of configuration over a precipice of more than one hundred feet, but presenting no single body equal to the Gungani Chuki, but the whole forming an amphitheatre of cataracts, meeting the eye in every direction along a sweep of perhaps 90°, and combined with scenery of such sequestered wildness that for picturesque effect it is perhaps without parallel in the world." This branch of the stream is used to irrigate the province of Tanjore, and the coming of its floods is celebrated by the natives with special festivities, as they consider the river to be one of their most beneficent deities.

The beautiful and picturesque fall of the Rhine below Schaffhausen, where the water falls sixty-five feet in a single column, is the admiration of all travelers.

Victoria Falls—Zambesi.

CHAPTER XXI.

Famous Rapids and Cascades — Niagara — Amazon — Orinoco— Parana — Nile — Livingstone.

IN all its features and characteristics the great watercourse, including the great lakes, which feeds the Niagara, is peculiar and interesting. It is more than two thousand miles long; its utmost surface-sources are scarcely six hundred feet above tide-water; its bottom, at its greater depth, is more than four hundred feet below tide-water. In all its course it receives less than two score of affluents, and only two of these, the St. Maurice and the Saugeen, bring to it any considerable quantity of water, and no flood in any of them discolors its emerald surface from shore to shore. Only fierce gales of wind bring up from its own depths the sediment that can discolor its whole face. Far the greater portion of its water-supply is drawn from countless hidden springs, lying deep in the bosom of the earth. In all the elements of beautiful, picturesque, and enchanting scenery it is unrivaled.

The rapids of the Niagara just above the Falls, from the Leaping Rock down through the Witches' Caldron to the edge of the precipice, are nearly a mile in width, and discharge ten million cubic feet of water each minute. But for a combination of grandeur and beauty, and for imparting a sense of almost infinite power, nothing can surpass the Whirlpool Rapids below the Falls, where the ten million cubic feet of water are compressed into a tortuous, tumultuous channel, less than four hundred feet wide.

There are many lesser rapids in the St. Lawrence, from the Thousand Islands to Montreal, the passage of which in the large lake steamers is an exciting voyage. The constant changes of scenery at every turn and in every rood of progress is almost bewildering. Then the alternation of rapids and broad expanses of river, the bird-like motion as the steamer sinks and sails down through the rapids, and the sense of relief when it seems to rise and glide over the smooth river, vary and increase the excitement. There is developed in one of those expanses a peculiar geological feature called the Split Rock. The name is strictly accurate. The descending steamer finds but one narrow channel, a little more than its own width, through which it can pass in a stream more than half a mile wide. It lies between the sharp corners of a broad, wedge-shaped cleavage in an immense rock which, by some convulsion of nature—not by any abrading process of the elements—has been literally split downward more than eighty feet. The last crooked and turbulent rapid passed just before reaching Montreal is the terror of the river pilots, and they never attempt its passage except by daylight. From Montreal to the Gulf of St. Lawrence the constantly deepening channel flows with an unbroken current.

It is a notable fact that the great river of rivers, which drains a larger territory than any other on the globe, the Amazon proper, has a fall of only two hundred and ten feet in a course of three thousand miles, and while it has a deep channel and a uniform current of three miles an hour for its whole length, it has no broken rapids. But in its many great affluents rapids are numerous, though not so famous as those found in other South American rivers.

The river Orinoco, more remarkable in some respects than the Amazon, receives the waters of four hundred and thirty-six rivers, besides two thousand smaller streams. It is one thousand five hundred miles long, is navigable for seven hundred and eighty miles, and at Bolivar, two hundred and fifty miles from its mouth, it is four miles wide and three hundred and ninety feet deep. Its famous rapids of the Apure and Maypure were visited by Humboldt. At the latter, the river is two thousand eight hundred and forty yards wide, and plunges down an inclined plane about three miles long, making a fall equal to forty feet in vertical height. It is dotted with innumerable islands which furnish a striking contrast to the vast sheet of white water, presenting the singular appearance of an eruption of shrub-crowned rocks in a sea of foam. These islands, and its great width, constitute the peculiar characteristics of this chute.

In the grandest of the South American rapids, those of the river Parana, a vast volume of water from a channel nearly two and a half miles in width is compressed into a gorge only sixty-six yards wide, through which the flood dashes down a slope of sixty degrees inclination and fifty-six feet perpendicular fall. Its roar—a perpetual monotone—is heard thirty miles away.

Hardly less remarkable than the rapids of the South American rivers are those of the two great African rivers, the Nile and the Congo, or, as Mr. Stanley has re-christened the latter, the Livingstone. The Nile may be compared to a vast tree with its huge delta-roots in the Mediterranean, its boll extending up through a rainless desert nearly one thousand five hundred miles to meet its numerous branches which stretch up into the mountains of Abyssinia, and the vast basin south of the equator that contains the great lakes of Victoria N'yanzi and Albert N'yanzi. From these branches in each year, at a fixed season, are poured down the sediment-charged waters which irrigate and fertilize an immense valley that would otherwise be only a parched and desert waste.

Without specifying the data for his calculations, Mr. Stanley, who saw them both, states that the volume of the Livingstone is ten times greater than that of the Nile. Its course is interrupted by two series of cataracts, or rather a combination of cascades and rapids. The first series, seven in number, occurs within four hundred miles of its source, and consists of the Stanley Falls, occupying different points in a channel sixty-two miles long. Its banks are of moderate elevation above its bed, and in the long, bright, equatorial days the leaping, sparkling, foaming waters present a scene of dazzling brilliancy. In the second series, named by Mr. Stanley the Livingstone Falls, there are thirty-two cascades, more extensive and imposing than those of the first. The river, after a gentle descent of nearly one thousand miles, and after receiving many large affluents, reaches the first of these impetuous torrents where all its waters are compressed into a narrow gorge only four hundred and fifty feet wide, and at a single point near the right bank where a sounding was possible, Mr. Stanley found a depth of one hundred and thirty-eight feet.

The remaining thirty-one cascades are distributed along a channel one hundred and fifty-five miles in length, between banks from fifty to six hundred feet high, and having a fall of one thousand one hundred feet. The dimensions here given indicate that these rapids are second, in power and impressiveness, only to those above the Whirlpool of Niagara.

www.ingramcontent.com/pod-product-compliance
Lightning Source LLC
Chambersburg PA
CBHW021410230426
43666CB00006B/701